THE
ATLAS
OF
WILD
PLACES

THE
ATLAS
OF
WILD
PLACES

*In search
of the
Earth's
last
wildernesses*

ROGER FEW

*Foreword by
Dr. Thomas E. Lovejoy
Smithsonian Institution*

SMITHSONIAN BOOKS
WASHINGTON, D.C.

A Marshall Edition
This book was conceived, edited,
and designed by Marshall Editions,
170 Piccadilly, London W1V 9DD

Smithsonian Books, an imprint of
Smithsonian Institution Press
470 L'Enfant Plaza
Suite 7100
Washington, D.C. 20560

ISBN 0-89599-047-4

Contributors
Roger Few Pages 6–7, 10–29, 66–97,
98–131, 156–187, 188–199
Duncan Brewer Pages 132–155, 226–236
Sophie Campbell Pages 200–225
Ed Zahniser Pages 30–65

Editor   **Jinny Johnson**
Art director   **Dave Goodman**
Picture editor   **Zilda Tandy**
Research   **Jon Richards**

Printed in Spain by Printer Industria Grafica

10 9 8 7 6 5 4 3 2 1

This book is printed on acid-free paper

Page 1
A huge salt flat on the Bolivian Altiplano
Page 2
Peyto Lake in Banff National Park

CONTENTS

*Grizzly bears roam the tundra in Denali
National Park.*

Herds of lechwe antelope graze on the Okavango Delta.

FOREWORD

"Only in wilderness can we truly understand ourselves and our place in nature and the world"

In January 1989, I started down the trail to Camp 41 in the Brazilian Amazon for the first time in more than 18 months. With me were United States Senators Tim Wirth, the late John Heinz, and Al Gore, as well as Washington *Post* Executive Editor Ben Bradlee, *Jaws* author Peter Benchley, and many others. It was one of the few moments while I was guiding this Congressional delegation to look at Amazon deforestation that we were not trailed by television cameras.

For everyone but myself, this was the first direct experience of the incredible diversity of a tropical rainforest. For all of us, it was a riveting and bonding adventure. For me it was something else as well: I was hardly five steps into the warm, moist forest, alive with night sounds, when I realized that this wild place I had chosen for scientific research had become something more. I felt as if I were coming home.

There is something in all of us that yearns for wild places. Sometimes this feeling is suppressed by other interests and distractions. Sometimes it

Zebras in the grasslands of the Serengeti in Africa (above).

The vast Empty Quarter of Saudi Arabia (right).

remains in the abstract, as it was for me as a boy. I would stare for hours at a globe, relishing in imagination visits to places such as the Rub al-Khali, Saudi Arabia's fabled Empty Quarter; places the very names of which were to be savored.

There are considerably fewer wild places now than when I was a child. We have a way both deliberately and unconsciously of encroaching upon and diminishing them. The grand wilderness of the Florida Everglades, another of the objects of my youthful reveries, is so greatly altered that scarcely a drop of water in the great "river of grass" flows naturally any more.

This volume is a wonderful reminder of the value of wild places and a tribute to those that are left. I have had the good fortune to have visited some of these areas; those that I haven't seen still awaken in me the identical longing I

The desolate, icy wastes of the remote Antarctic Peninsula (above).

felt as a child, enthralled by their exotic names as they appeared on the globe.

Only in wilderness can we truly understand ourselves and our place in nature and the world. The reality is that most of us will not be able to experience wilderness firsthand, but we are lucky to have books such as this one to bring us closer by knowing if not by visiting. Nothing could be more important. Were the last wild places to be destroyed, I am certain that something both profound and fundamental in all of us would die.

Thomas E. Lovejoy

Dr. Thomas E. Lovejoy
Assistant Secretary for
Environmental and External Affairs
at the Smithsonian Institution

THE WORLD'S WILD PLACES

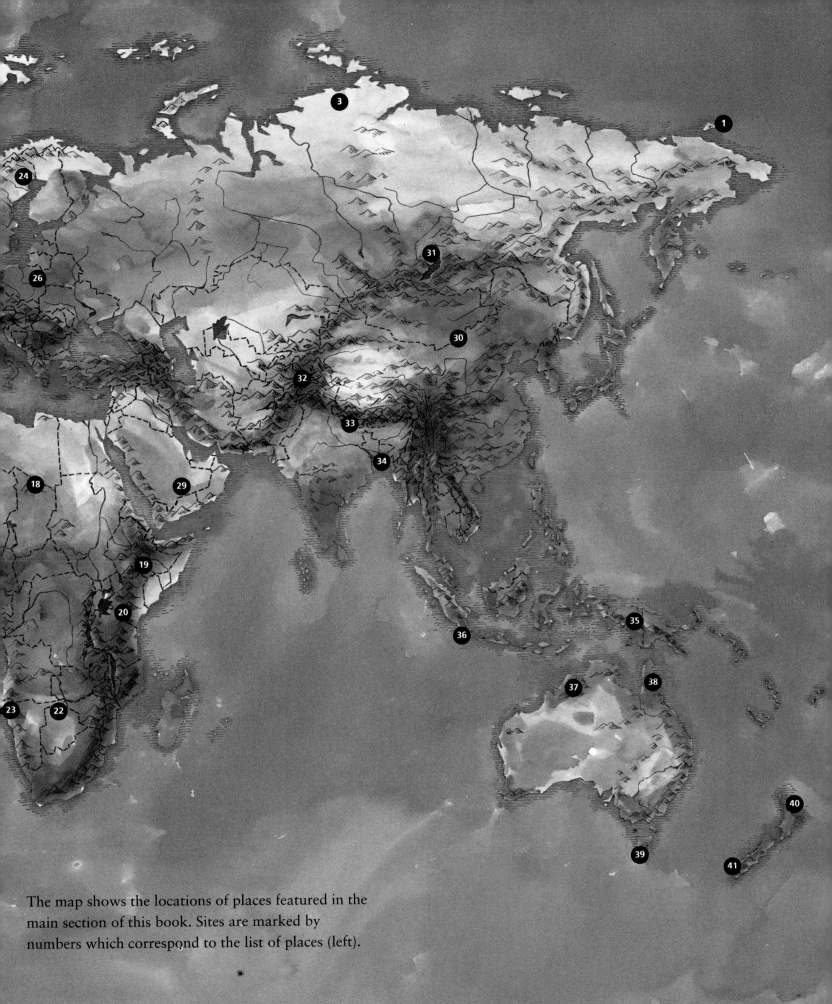

The map shows the locations of places featured in the main section of this book. Sites are marked by numbers which correspond to the list of places (left).

WRANGEL ISLAND

"An uninhabited polar wilderness, isolated from the rest of the world"

Pack ice

ARCTIC

WRANGEL
ISLAND

N

A remote outpost in the Arctic Ocean, Wrangel Island lies off the coast of Siberia. It is located near the limit of permanently frozen ocean surface, and approach by sea is made treacherous even in summer by shifting pack ice. In the cold dark winter, a platform of solid sea ice several miles wide seals the shores of this uninhabited polar wilderness, making it still more isolated from the rest of the world.

About 78 miles (125 km) in length, the island seems to guard the entrance to the Arctic above the narrow Bering Strait that divides Siberia from Alaska. Wrangel takes its name from a Russian explorer, who knew of its existence but never had the chance to glimpse its desolate shores. In the 1820s, while mapping the northeastern Siberian coast, Ferdinand Wrangel noted flocks of birds flying out and back across the icebound sea. Local Chukchi people confirmed the belief that there must be land somewhere to the north in the frozen wilderness.

But such was the isolation of the island that it was not definitely sighted by Russians until 1849. When explorers finally reached its shores, they found a pristine, remarkably ice-free land – a wild haven not just for huge numbers of migratory birds, but also for true Arctic wanderers such as walruses and polar bears.

Unlike many other islands of the High Arctic, Wrangel does not have any permanent ice cap nor any glaciers. Thin winter snow covers its undulating mountain chains and coastal lowlands, and deep, windblown drifts may accumulate

Remote Wrangel Island is a haven for wildlife. Lesser snow geese migrate to the island in the thousands in summer, finding safe nesting sites and plentiful food on the ice-free tundra.

in the valleys, but the relative warmth of summer is enough to melt away most of the snow each year.

During the brief Arctic summer, numerous rivulets and streams take meltwater down through ponds and marshes to the coast. The rugged tundra landscape through which these streams flow is painted mostly in shades of russet and brown with scant, patchy vegetation rarely growing more than about 4 inches (10 cm) in height. Among these diminutive plants are some flowering species – types of poppy, meadow grass and cinquefoil, for example – that are unique to the island.

The small size of the plants is both a result of the short growing season and an adaptation to the climate. Ground-hugging plants are less exposed to winter gales and are better protected from intense cold. The hardiest plants, such as mosses and lichens, can grow outside the valleys and lowlands, but even they cannot survive on the highest ground of the

Thin summer ice around Wrangel reflects mild conditions in comparison with the island's ice-age past. Its peaks, once part of a great mountain chain, were worn down and smoothed by thick ice sheets that covered the land.

island. There, not even the thinnest covering of soil exists, just shards of rock debris. These lie loose on the surface in summer, but are locked together by the permanent frost underneath.

Before the snows have fully melted on Wrangel Island, summer visitors are already winging into the sheltered grassy valleys and shoreline plains ready to begin nesting. These birds must arrive early if they are to complete their breeding cycle. By the end of August, worsening weather will have driven most of them away again. Though the trip across to Wrangel is too arduous for many species that nest in Siberia, those that do extend their journey north come in large

numbers. Especially numerous are brent geese, common eiders, knots, turnstones and gray plovers.

Wrangel is also the only place in this part of the Arctic where large colonies of lesser snow geese nest. By May these graceful white birds can be seen in the air all over the island, traveling to and from colonies shared by many thousands of pairs. Snowy owls often nest alongside the colonies in what seems to be a deliberate strategy for mutual protection against nest-robbers. The watchful geese provide early warning when marauding gulls, skuas and Arctic foxes approach, then the owls swoop in to drive the intruders away.

Along the rocky coasts of the island, birds gather in even greater concentrations. The best sites are crowded with nesting guillemots, cormorants, kittiwakes and other birds that constantly wheel in the sky around their cliff-ledge haunts. At least half a million seabirds throng Wrangel's coast at the height of the Arctic summer, finding abundant food in the surrounding waters as some ice thaws.

The break-up of the coastal ice also allows sea mammals to swim inshore, where the water is shallow and food easier to find. As soon as openings in the ice appear, so do bearded and ringed seals. By July walruses, too, are hauling themselves onto traditional beaches to court, spar, mate and raise their pups. Wrangel Island is one of the most important breeding areas for walruses in the world, in some years attracting as many as 80,000 of these hefty, tusked sea mammals.

Another spectacular animal, the polar bear, also breeds on Wrangel. Protected by its thick fur and layers of subcutaneous fat, the polar bear rides on ice floes and braves the freezing water for short swims as it searches for prey. Its principal victims are seals, which it catches on the ice or

waits to club with its paw when they surface at breathing holes.

By November, when winter returns to Wrangel, a few hundred bears will have made their solitary way inland to occupy lairs on snowy hillsides. Many are pregnant females, and during the midwinter darkness, they give birth to their cubs under an insulating blanket of snow. Not until April can the cubs emerge into the gentle spring sunshine.

A few thousand years ago, very different creatures roamed Wrangel Island – the wooly mammoths. These prehistoric, elephantlike beasts were generally believed to have died out worldwide at the end of the ice age, more than 10,000 years ago. By that time, rising sea levels had cut the island off from the mainland.

But in 1991, paleontologists found some much more recent remains on the island. These indicate that mammoths could have lingered undisturbed in this lonely wilderness until as late as 2000 B.C.

Up to 300 polar bears return winter after winter to denning areas in the interior of Wrangel Island. For the rest of the year, bears like this mother and her well-grown cubs (above) are wanderers on the polar pack ice.

A knot of guillemots finds space on a Wrangel cliff alongside a noisy colony of black-legged kittiwakes. The continual din from nesting seabirds shatters the peaceful desolation of the island in the summer months.

ELLESMERE ISLAND

"A truly forbidding place, one of the great wildernesses of polar lands"

Ellesmere Island's bleak, rugged interior is no place for the lonely. The most northerly of Canada's Arctic lands, it lies 2,500 miles (4,000 km) from Canada's main cities. Though it is the size of Great Britain, this beautiful but empty land is home to just a few hundred people, clustered into a handful of tiny settlements and research stations along the coast. Intense cold, scant vegetation and

the prolonged darkness of the Arctic winter combine to make the island a truly forbidding place, one of the great wildernesses of the polar lands.

Conditions on Ellesmere are governed by latitude. Straddling the 80th parallel, the island lies almost at the top of the world. Its northernmost point, Cape Columbia, is less than 500 miles (800 km) from the North Pole. Only the extreme tip of Greenland is nearer. At this latitude,

the seasonal effect on climate caused by the tilt of the Earth as it orbits the sun is dramatic. For most of winter, the sun never rises above the horizon. The land is plunged into continual darkness that lasts up to five months in the north of the island. For a similar period in summer, the sun remains in the sky, dipping daily toward the horizon, but never below it.

Long months without sunshine make the Ellesmere winter exceptionally

Ellesmere is the most northerly and mountainous of the Canadian Arctic islands. Ice caps cover the higher ground, and for much of the year snow whitens the entire landscape.

cold – average midwinter temperatures are well below -22°F (-30°C). Even at midday in summer, the sun is so low in the sky that it does little more than take the chill off the air. Average

summer temperatures on the island are only just above freezing.

Year-round pack ice (frozen surface layers of the sea) fills the Arctic Ocean to the north and the straits and channels to the west. Only to the south and in the narrow strait separating Greenland to the east does the coastal water become clear for shipping in summer. Inland, huge sections of higher ground are covered in permanent ice caps. Snow whitens the ice-free land, the arctic tundra, for nine months each year. Ironically, however, Ellesmere's environment is classed as desert. Over most of the island, only about 2½ in (65 mm) of snow falls per year – enough to coat the ground, but less moisture than the Sahara Desert receives yearly as rain.

Survival on Ellesmere Island is difficult. Archaeologists have found traces of ancient Inuit (Eskimo) settlements on the island and even evidence that there may have been some trading with Viking adventurers in the 12th century. The island probably served as a migration route between islands to the southwest and Greenland. But until the tiny modern Inuit settlement of Grise Fjord was

founded on the south coast in 1953, nobody had made a permanent home on the island for 250 years. The only visitors were Inuit hunters and a few scientists and explorers.

The fate of one expedition underlines the harshness of Ellesmere. In 1883, having failed to receive relief supplies by ship for two years running, an American expedition party which had been exploring the northern part of the island retreated south in desperation, amid gales, ice and blizzards. When they reached the emergency stores they were seeking, they found only meager supplies with yet another terrible winter to face abandoned. By the time they were rescued the following summer, only 7 of the original 25 had survived.

Harsh though the dark winter may be, Ellesmere in the light of summer presents a majestic face. As the light strengthens and the snow thaws, the landscape revealed is one of towering mountains up to 8,500 feet (2,600 m) high, vast ice caps, glaciers, rugged stony plateaus, and an emerging flush of green in the valleys. Magnificent fiords indent the coast, and icebergs drift menacingly offshore.

Peary's caribou is the smallest of all the races of caribou and one of the island's most enchanting inhabitants. But life on Ellesmere is tough for these deer, forcing them to wander far afield in search of meager vegetation, which they expose by scraping the snow aside with their hooves.

The landscape is one of vast, open tundra. No trees can exist here – only dwarf willow shrubs manage to survive the rigors of winter. Various lichens, mosses, grasses and small herbs such as saxifrages and Arctic poppies, however, relieve the bleakness of the tundra and provide a lifeline for wildlife. In sheltered valleys the plants cover the ground thickly, creating pockets of relative plenty.

On the richest parts of the tundra, animal life appears in summer – if not in the variety present in warmer climes, then in an abundance that may still seem hard to reconcile with the harshness of winter. But most creatures have done their best to avoid the Arctic winter. Some come out of hiding; others arrive from more clement winter quarters. Hardy

"Harsh though the dark winter may be, Ellesmere in the light of summer presents a majestic face"

VERDANT VALLEYS

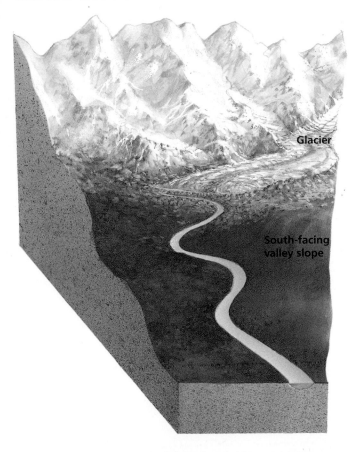

Glacier

South-facing valley slope

On clear summer days, sheltered valleys on Ellesmere can become oases of warmth in a chilled land, with temperatures rising from time to time as high as 68°F (20°C). The sun is continually low on the horizon, and south-facing slopes become effective suntraps. Snow melts away quickly on them, and pockets of surprisingly lush greenery can develop.

insects, among them butterflies, emerge from winter dormancy in the soil and mosses. Great numbers of birds, such as plovers, sandpipers, ducks, geese, waders, terns, auks and skuas, fly in from lands far beyond the Arctic to nest and feed in the tundra, along rivers and lakes, and on the coast. Shortly after breeding is complete, they retreat south again.

Only the few mammal species resident on the island remain active as winter sets in. Lemmings, Arctic hares and Arctic foxes do not hibernate – if they did they would probably freeze to death. As the blizzards start to swirl, caribou paw snow aside to find forage. When winter darkness falls, life is grim indeed.

Musk oxen, protected by shaggy coats, move up onto the upland plains, where at least the strong winds blow some of the snow cover away and allow them to find morsels of plant food. Arctic wolves may rely a good deal on scavenging. They trek far and wide in the gloom to find the bodies of musk oxen and other mammals that have succumbed to the severity of winter in this wild island of the High Arctic.

Arctic poppies prepare to unfurl their flowers in the precious warmth of summer, adding their delicate beauty to the harsh grandeur of Ellesmere's tundra. The petals attract hardy insects emerging from winter dormancy, though they may also suffer the nibblings of Arctic hares.

THE ARCTIC WOLVES OF ELLESMERE

In the wilderness of Ellesmere Island, far from the prejudice and persecution meted out to them in settled lands, Arctic wolves hunt and breed without menace from the gun.

But life is by no means easy. The thick white fur of the wolf is testament to a landscape gripped by intense cold, and though the diet is broad – wolves hunt or scavenge for musk oxen, caribou, lemmings and ground-nesting birds like the ptarmigan – prey is so thin on the ground that each pack needs a territory of perhaps 1,000 sq. miles (2,600 km^2) if it is to find enough food.

Leaping between drifting ice floes, a male Arctic wolf patrols the edge of his territory (left). Coastal areas are valuable to a wolf pack, providing plenty of opportunities for scavenging and the chance to hunt for resting seals.

Musk oxen form a defensive ring when attacked, with the adults' horns facing menacingly outward (above).

Adults in the pack share the responsibility of looking after pups (right).

THE TAYMYR PENINSULA

"Views can be breathtaking, especially in winter when the northern lights illuminate the dark snow and frozen waterways"

Bordered by ice-filled seas, raked by terrible blizzards and plunged each year into four months of perpetual darkness, the Taymyr Peninsula reaches farther into the Arctic than any other part of Siberia. Cape Chelyuskin, its northernmost point, is the closest any continent gets to the North Pole. It is a part of Russia that has long deterred even the hardiest of settlers, and its coastal region was one of the last in Siberia to be explored.

Still virtually devoid of people, Taymyr has enormous vistas of open

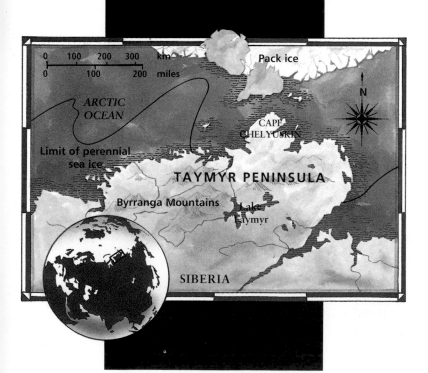

0 100 200 300 km
0 100 200 miles

Pack ice

N

ARCTIC OCEAN

CAPE CHELYUSKIN

Limit of perennial sea ice

TAYMYR PENINSULA

Byrranga Mountains

Lake Taymyr

SIBERIA

tundra, crossed and dotted by myriad rivers, pools and lakes. Its views can be breathtaking, especially in winter when the atmospheric extravaganza of the northern lights, or aurora borealis, delicately illuminates the dark snow and frozen waterways.

Tundra stretches across the entire northern rim of Russia, sandwiched between the great Siberian forests and the Arctic coast. At Taymyr, with its extension deep into the Arctic, this tundra belt assumes its greatest width, forming an enormous block of treeless, windswept land the size of

Germany. The peninsula is largely flat or undulating, although a ridge of low, worn mountains, the Byrranga, crosses its center and in places forms precipitous crags.

As throughout the Arctic, the ground underfoot is frozen solid by permafrost – subsoil that remains below freezing point throughout the year. In this region of Siberia, probably because of the severity of past glaciation, the deep freeze penetrates far underground, as much as 1,000 feet (300 m) into the subterranean rock strata. In winter,

when temperatures can plunge below –58°F (–50°C) at times and gales can drive snow across the land for days on end, the entire landscape of Taymyr lies frigid and white. But summer sunshine not only melts away most of the snow, but also warms the surface of the soil enough to thaw the top 4 inches (10 cm) or so of frost.

Without this annual melting, the short tundra vegetation would not be able to survive and grow. In fact, compared with the grip of winter, summer on Taymyr is surprisingly warm, averaging 46°F (8°C) in July and rising to an exceptional 86°F (30°C) in places on still, sunny days.

The great latitudinal span of the Taymyr Peninsula creates varying conditions from north to south. In the far north, where the climate is coldest, vegetation is scant and there are large patches of bare ground. Only the hardiest Arctic animals eke out an existence here. The most numerous are those typical Arctic rodents, the lemmings. Two kinds live on the peninsula, the collared lemming and the Siberian lemming, digging holes in the snow for shelter during the icy winter.

Farther south, across the Byrranga Mountains, the ground is more thickly covered with mosses, lichens, sedges, grasses and dwarf shrubs. Particularly abundant is "reindeer moss" – actually a type of branching lichen that grows in tight mats up to 6 inches (15 cm) high. As its name suggests, the moss is a staple food of reindeer or caribou, which spread north across the peninsula in summer

The lonely Taymyr Peninsula lies farther north than any other part of mainland Russia. Tundra and shallow summer water stretch for immense distances across its open terrain. Apart from a line of low mountains crossing its center, there is little in this vast land to break the horizon.

in herds thousands strong. Up to half a million of them may be present in all, making Taymyr one of their main strongholds in Siberia. By September, each herd begins to trek south again, often shadowed by a pack of wolves.

Along their route south, the surrounding vegetation becomes steadily thicker as the moss tundra blends into the zone of shrub tundra. The soil becomes a little deeper and more peaty, and dwarf forms of birch and alder – 3 feet (1 m) in height – form thickets over large areas. They provide good nesting sites for small tundra birds such as snow buntings which, when they appear in early May, are among the first of the summer migrants to the peninsula.

Everywhere, especially in this southern half of the peninsula, water seems to be abundant. Melting snow and soil frost swell the winding rivers and shallow lakes. At this time, Lake

Every spring, *once the winter snows have thawed, herds of caribou or reindeer trek deep into Taymyr to find fresh feeding pastures. Here they fatten themselves on the lush tundra vegetation, a diet complemented by birds' eggs and the occasional lemming. Females calve in the relative warmth of June, each rearing a single young that is up and running within hours of birth.*

Taymyr in the center of the peninsula gradually doubles in surface area to 1,780 sq. miles (4,600 km²) though it is nowhere deeper than 85 feet (26 m). Over flat areas, the soil holds the meltwater like a sponge, since permafrost prevents the water from percolating downward, and the ground becomes sodden and marshy.

One of the characteristic effects of permafrost in tundra zones is to pattern the ground. Complex cycles of freezing and thawing heave up the soil, move and sort debris, and create angular fissures. On level terrain, they make the ground seem drawn into polygonal sections, sometimes disarmingly regular as if traced by giant human hands. The polygon rims tend to be higher than their centers, allowing ponds that accentuate the patterns to form in each.

All this fresh water and marshland, bordered by bright-flowered meadows, attracts untold numbers of nesting birds. During the warm days of continual sunshine, they can forage at any time. Gulls and terns glimmer white in the air, at times twisting and turning as piratical skuas harass them to relinquish their food. Many spectacular fowl nest close to the water, among them white-billed divers, Bewick's swans, king eiders and long-tailed ducks. The most beautiful is surely the red-breasted goose, a threatened species for which the peninsula is the single most important breeding area.

By the time the last migratory birds have left, the caribou, too, are well advanced on their terrestrial migration. Some linger through winter on the tundra, but most head away from the peninsula deep into

The red-breasted goose breeds on Taymyr. Small flocks of the geese appear at ice-free lakes and rivers in June, ready to make nests on mounds and clumps of vegetation.

the forest margins, where food is easier to come by. At the southern limit of the peninsula, stunted larches about 10 feet (3 m) high appear. These pockets of trees, surrounded by open tundra, are the northernmost forests anywhere on the planet.

TUNDRA PERMAFROST

Deep permafrost benumbs the tundra soil, but in the brief summer the surface thaws enough for plants to grow in this "active layer." Because surface water cannot penetrate the permafrost and drain away, the ground becomes marshy in summer and pools develop.

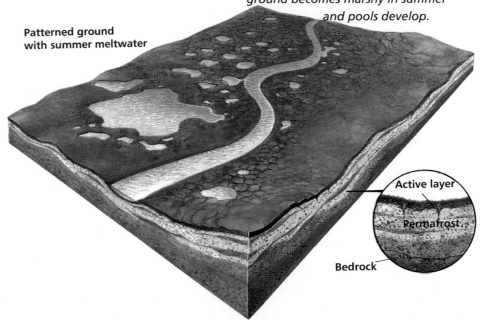

Patterned ground with summer meltwater

Active layer

Permafrost

Bedrock

THE ANTARCTIC PENINSULA

"Life has a firmer foothold here than elsewhere on Antarctica"

Antarctica has the harshest climate of all of the Earth's wildernesses – it is colder, drier and windier here than anywhere else in the world. But one part, the Antarctic Peninsula, which stretches out across the Antarctic Circle to within 600 miles (1,000 km) of South America, has a slightly less severe climate than the continental heartland. Surrounded by the

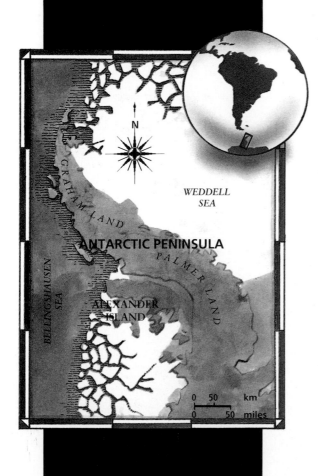

GRAHAM LAND

WEDDELL SEA

ANTARCTIC PENINSULA

PALMER LAND

BELLINGSHAUSEN SEA

ALEXANDER ISLAND

N

0   50   km

0   50   miles

moderating influence of the ocean, the peninsula is a little warmer and more humid than the rest of the continent and, as a result, plant and animal life has a firmer foothold here.

The backbone of the Antarctic Peninsula is a chain of mountains that begin to rise in western Antarctica. They run the entire crooked length of the peninsula, raising it to heights of nearly 11,500 feet (3,500 m) in the loftiest sections. Geologists regard these mountains as an extension of the South American Andes, and it has been suggested that a land bridge once connected the two continents.

Palmer Land, the poleward section of the peninsula, is a curving finger of land up to 155 miles (250 km) wide between the Bellingshausen and Weddell seas. The ice sheets that cover most of its interior form a

On the wild and desolate Antarctic Peninsula, realms of mountain, ice and ocean converge. Here, where exposed rock interweaves with icebound terrain and snows melt briefly in summer, a few species of plant and animal struggle to survive.

AN OCEANIC CLIMATE

Like a giant spur on the fringe of Antarctica, the peninsula thrusts out into the oceanic zone. Its climate is made slightly warmer and moister than that of the rest of the continent by the heat-retaining effect of the ocean and the greater humidity of the sea air.

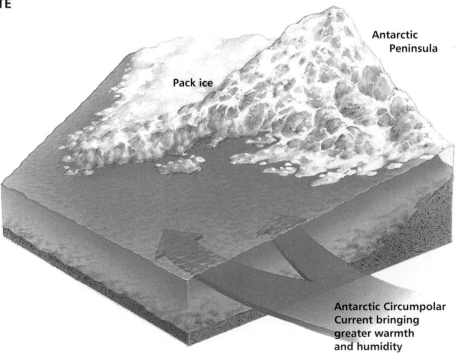

Pack ice

Antarctic Peninsula

Antarctic Circumpolar Current bringing greater warmth and humidity

Treacherous sea ice forms around the Antarctic Peninsula during the long and bitter polar winter (above). These frozen rafts of sea water form vast, unstable plains that are interrupted only by mountainous floating icebergs.

plateau more than 6,500 feet (2,000 m) high. Graham Land, the northern section juts from it like a recurved talon. Only 25 miles (40 km) wide in places, it rises from the sea in steep cliffs leading to snowbound peaks. Glaciers flowing from the mountains to the sea have sculpted deep valleys and indented the coastline with fiords.

Islands flank both coasts of the peninsula, separated from it by deep straits and channels. Many, like the giant Alexander Island west of Palmer Land, are locked for most of the year within huge ice shelves (floating platforms of glacier ice) or pack ice (frozen surface layers of the sea). Only the western shore of Graham Land is relatively free of sea ice through the Antarctic winter.

Nevertheless, the peninsula's icy climate is the mildest in Antarctica – though its nickname "the banana belt" stretches the point too far. In the north, temperatures may creep above freezing point even in winter, and it can be 59°F (15°C) in the peninsula on a midsummer's day. The average for the year is about 27°F (–3°C) compared with an average of –58°F (–50°C) at the South Pole. The peninsula also receives much more precipitation than the rest of Antarctica, which is classed as desert. Most of the continent receives only 2 inches (50 mm) of moisture in the form of snow in a year, but the peninsula has ten times this amount.

Such conditions make it possible for two species of flowering plant to survive here – the only ones known in Antarctica. But in general, the dominant plants are diminutive, hardy kinds. Cushions of moss and liverworts appear on ice-free rocks along the shore, and tiny fungi lace the pockets of soil. Smears of algae can even spread across snow and ice, staining it green or red with their pigmentation.

The most abundant and successful plants of all are the lichens. About 350 kinds occur in Antarctica, well adapted to the extremes of climate. Most ignore the rigors of winter by becoming dormant, closing down the processes of life to conserve their resources and prevent damage by freezing. But in summer they begin photosynthesizing, making the most of the slanted sunshine. They plaster exposed rocks all over the continent, even some isolated peaks that poke above the ice sheet close to the pole. Though many are black or gray, others are bright orange and green, bringing splashes of color into the peninsula landscape.

The dwarf "forest" of Antarctic plants is home to wildlife of a similar scale. The only truly terrestrial animals that can cope with Antarctica

Only two types of flowering plant grow in the whole of Antarctica – the Antarctic hair grass and the pearlwort. Both grow near the coast on Graham Land, usually on rocky, north-facing slopes from which snow thaws more easily in summer. The grass tends to grow in clumps, but occasionally forms a closed turf over the surface.

are either microscopic or of small insect size. They include protozoa, rotifers, roundworms, lice, fleas and gnats. Mites and springtails barely a millimeter in length cluster on plants and under rocks and forage on fungi and algae or, in some cases, on each other. The predatory mite *Rhagidia gerlachei* eats adult springtails and the

Gentoo penguins bask in the muted light of the Antarctic sun. Like Adélie and chinstrap penguins, gentoos nest on rocky and tussock-covered slopes on the peninsula, but in winter they move out beyond the expanding margins of the sea ice.

eggs, and is one of many creatures that zoologists suspect have a form of antifreeze in their body fluids – a substance that stops them from freezing rigid at winter temperatures as low as –30°F (–34°C).

Bigger animals are not permanent residents in Antarctica. They may use it as a base for resting and nesting in summer months, but spend most of their time either in or over the ocean. Seals swim in the waters off the peninsula and haul up onto the beaches in summer to give birth and rear their pups. Millions of birds, including petrels, gulls, terns and blue-eyed cormorants, flock to nesting colonies on the rocky coasts

and islands. Rich waters full of krill and fish provide them with plenty of food for their young. Four kinds of penguins breed in the area, and brown skuas and snowy sheathbills lurk among the colonies ready to steal eggs and chicks.

Most birds move away – even from the peninsula – in winter, but the emperor penguin holds out against adversity. Colonies of these birds, which are 3 feet (1 m) tall, huddle together on the ice fields of Palmer Land through the long, dark winters. Males take the first three-month shift, each bird incubating a single egg between its feet and its abdomen, followed by the females,

which return from the sea when the chicks are about to hatch. The male emperors then make the long journey across the ice to feed in open water.

There has long been a sparse human presence in Antarctica. Sealers operated in this region for most of the 19th century, but the greatest impact came with intensive whaling in the 20th century. Factory ships moved into the sheltered waters, processing stations were built on the shores, and harpoon gunners soon brought about a collapse in the great whale population. Though the industry has now been banished, the blue whale, which was hunted almost to extinction in the southern oceans, has

"The only true land animals that survive on Antarctica are of microscopic size"

scarcely begun to recover. The most likely cetaceans to be seen today around the peninsula are killer whales cruising offshore, waiting to snatch penguins and seals.

Now the sealers and whalers have gone, but the scientists have moved in. Research stations and bases are concentrated on the peninsula, the most hospitable and accessible part of

the continent, representing 10 nations in all. These bases reflect the desire of many countries to establish an official presence in – and therefore a stake in the future of – Antarctica. Some have already brought pollution to the fragile polar continent, but for now at least, this barren but beautiful land remains one of the greatest of all wildernesses on Earth.

DENALI NATIONAL PARK

"Mount McKinley, North America's highest peak, crowns Denali's rocky majesty"

A magnificent expanse of subarctic wilderness, the Denali National Park and Preserve spreads its stark mantle of pristine snow and jutting rock over an area larger than the state of Massachusetts. Across this vast virgin landscape, which covers more than 6 million acres (2.5 million hectares) of the interior of Alaska,

The foothills of Mount McKinley stand as a massive backdrop to the wandering rivers and vast expanses of heathlike tundra of Denali.

roam herds of migrating caribou and Dall sheep, pinpricks of life in the enormous emptiness. And Mount McKinley, North America's highest peak and perpetually shrouded in snow, crowns its rocky majesty.

Athabascan Indians knew the mountain as Denali, "the High One." It forms the apex of the awesome 600-mile (1,000-km) Alaska Range, a natural barrier that separates Alaska's rugged interior to the north from the coastal lowlands to the south. The bulky, looming range dwarfs the surrounding panorama of treeless tundra and sparsely wooded taiga, threaded with braided, meandering rivers that originate in its own glaciers. The south peak of Mount McKinley juts higher than any other point in North America, rising 20,340 feet (6,200 m) above sea level in an astounding vertical sweep. In addition, its north summit ranks as North America's second-highest peak at 19,455 feet (5,930 m).

Solid ice hundreds of yards thick blankets the granite and slate core of Mount McKinley, and the permanent ice fields that cover more than half

In the vast wilderness of Denali, *grizzly bears roam far and wide in search of plants and berries to eat as well as prey. A full-grown grizzly stands up to 7 feet (2 m) tall and weighs up to 850 pounds (390 kg).*

"During the brief summer season, the rocks burst into glowing color as flowers bloom in amazing profusion"

the mountain feed a myriad of glaciers, which surround its base. Even in summer, temperatures are severe; in winter, temperatures on the mountain plummet below -94°F (-70°C) and slicing winds gust in excess of 150 mph (240 km/h).

Both of Mount McKinley's peaks, located on the Denali Fault which is the largest break in the crustal plate in all of North America, are still rising. Geologists now think that Denali may encompass up to 7 blocks of unrelated land, while Alaska itself may be composed of as many as 32 of these crustal blocks called terranes. Denali's seven terranes formed independently, in different places around the globe, but over eons of time the Earth's tectonic plate movements drove them together.

The arrival here, 100 million years ago, of one such block, made of volcanic islands, may have given rise to the Alaska Range, whose geological faults and folded rock

formations provide the evidence of its ancient birth. The monumental collisions and grindings would have produced spectacularly violent earthquakes and volcanic eruptions. Geologists now know that the pressure to fold these rocks into such high and bulky mountains came from the Kula Plate pushing beneath Alaska for 100 million years.

As a result of these thunderous birth pangs, Denali's wildlands display a fabulous rainbow of rocks. The Polychrome Pass, from which grizzly bears are often sighted, takes its name from the bright palette of colors of its folded rock. Cathedral Mountain and other lesser peaks are reddish, and Igloo Mountain, which is ice-free in summer, has masses of loose rock fragments, or talus, that display a wide range of hues.

After the continental ice sheets retreated about 10,000 years ago, centuries passed as a fragile layer of topsoil gradually accumulated and the slow process of revegetation began. Above the permafrost – permanent subsurface ice that has remained frozen for thousands of years – this thin cover of soil thaws just enough in summer to support successful plant growth. Two major types of plant community now exist in Denali – tundra and taiga.

Tundra may be moist or dry. On moist tundra, tussocks of sedges and cottongrass or dwarfed shrubs, such as birches and willows, carpet the ground. Dry-tundra plants occur in patches, sometimes dotting rocks at higher elevations. During the brief summer season, the rocks burst into glowing color as an amazing profusion of blossom attracts the insects that will help the plants reproduce before winter returns.

Taiga means "land of little sticks" in Russian, and the name aptly characterizes the tree growth close to the Arctic Circle. Denali's taiga, which mainly clings to the river valleys, is composed largely of white and black spruce trees, interspersed with quaking aspen, alder, balsam poplar and the white, or paper, birch. Open areas of taiga fill with shrubs – blueberry, willow and dwarf birch. In these desolate wildlands, the limit of tree growth occurs just 2,700 feet (820 m) above sea level, nearly 18,000 feet (5,500 m) below Mount McKinley's summit.

Much of Mount McKinley's character remains a secret of the wilderness, since the mountain is often wreathed in cloud. On a few summer days, the clouds briefly lift or part to reveal its hulking, snow-clad shape, perhaps still trailing enshrouding streamers of mist.

By contrast, the open, rolling lowlands of Denali National Park and Preserve seem tame at first sight. Shallow-bedded and laden with glacial flour – the liquid suspension of rock pulverized to powder by the grinding ice masses as they flow over the land – the rivers wander across the broad, flat valleys at will. They can dam themselves in a matter of hours with their own sedimentary load and sometimes set new channels overnight. All may appear calm, yet a grizzly bear may suddenly disrupt the tranquil scene, revealing the wild

The tundra in the fall glows with low-growing but brilliantly colorful foliage. Dwarf shrubs such as bunchberry, bearberry and Labrador tea (below) create swaths of red and gold across the land. In their brief growing season, these plants provide food for a multitude of tiny voles and rodents as well as the many birds that migrate north to nest in Denali.

elements in the sprawling landscape as it nonchalantly turns over huge chunks of tundra to lift the roof off the Arctic ground squirrel's burrow.

The grizzly's prey is also known as the parka or "parky" squirrel because the Indians of the Alaskan interior used its pelts to make winter parkas and lined the hoods with the pelts of long-haired wolves. But in their private wilderness drama, squirrels and wolves meet on different terms. While a wolf would rather make a feast of a Dall sheep or caribou calf, it often chases down the plump ground squirrel for a quick meal. Wolves may hunt alone in summer; in winter, they generally hunt in packs and often prey on moose, whose escape is hampered by deep snow in the willow thickets where they graze.

When the tundra and taiga are snowclad, little life moves across the severe landscape. Exceptions are ptarmigans, well camouflaged by their white winter plumage, as well as snowshoe hares and the owls and lynxes that hunt them. Slow-moving porcupines remain active, too, protected from most predators by their coat of quills. Occasionally, a lithe weasel will burrow under the snow and kill a porcupine by attacking its vulnerable underbelly.

Every summer, large populations of migrating birds cover the vast distances from Siberia, Central America and South America to breed in Denali. Many of these summer visitors feed on the hordes of mosquitoes and other flying insects that spring from the temporary fecundity of the moist tundra. Like subarctic plants, these insects must mature and reproduce quickly before winter clamps its deadly cold on them once again.

But the grizzlies, so called because of the grizzled appearance of the gray flecked coat, are perhaps the most potent symbols of Denali's wild seclusion. Hundreds of these bears, among North America's most formidable animals, inhabit the park.

Like the frozen river that it is, the huge Peters Glacier gathers tributaries as it carves a path from its source deep in the Alaska Range. But a glacier's tributaries do not merge into one; they remain separate, as the lengthwise striations here reveal.

Spreading antlers drip water as this moose, which has been feeding on succulent underwater plants, surfaces to keep a watchful eye on its surroundings. Moose also browse on dwarf willows and other low-growing tundra shrubs.

They have no natural enemies – only humans unnaturally armed with high-powered rifles can destroy them. Although bears, like cats and dogs, belong to the carnivore group of mammals, they tend increasingly toward omnivorous eating habits. The grizzlies in Denali subsist on berries and small plants such as vetch, as well as on ground squirrels, the calves of moose or caribou, and occasional carrion – the decaying carcasses of animals that have been killed or that have died as a result of disease or old age.

Grizzlies usually hibernate through the harsh winter, and females give birth to their cubs in underground dens during this period. When they first emerge into the dazzling spring sunshine with their cubs, female grizzly bears are at their most dangerous, prepared to defend their young against all enemies.

Despite their great size, ferocity and agility, grizzly bears do not hold absolute power over Denali's landscape. A full-grown Alaskan moose will stand its ground against a hungry grizzly, and a female protecting her calf will chase the bear away. The largest members of the deer family, Alaska moose can weigh 1,800 pounds (820 kg) and stand more than 7½ feet (2.3 m) high at the shoulder; a bull moose's antlers alone may spread 6½ feet (2 m) and weigh an incredible 90 pounds (40 kg). So even a predator as feared as the grizzly bear must exercise extreme caution to avoid injury.

Stunning contrasts heighten the thrilling impact of Denali: grizzlies feed on the tiny vetch, Mount McKinley's cloud-splitting peak towers over a lowland of vulnerable plants. And if Mount McKinley symbolizes the permanence of Denali's wilderness world, the youthful rivers and leaping streams exemplify its renewable vitality. Little altered by human intervention, here is a self-determined wilderness that retains its natural integrity.

Banff & Jasper National Parks

"Shining mountains rise above the valleys in matchless grandeur"

On the crest of the Canadian Rocky Mountains, the alpine peaks of Banff and Jasper National Parks pierce the sky at every point of the compass. Shining, skyscraping mountain walls rise above the valleys in matchless grandeur, like immovable bulwarks of nature damming the horizon. These

A breathtaking sunset over the snowcapped peak of Banff's Mount Rundle is dramatically mirrored in clear lake waters.

spectacular, adjoining parks straddle the border between Alberta and British Columbia in western Canada and between them cover a total area of some 6,760 sq. miles (17,500 km²).

Sediments of shale, dolomite, sandstone, limestone, quartzite and slate fuse here in the Canadian Rockies to form North America's upper spine, a skeleton so awesome that the Colorado Rocky Mountains seem puny by comparison. West of the Canadian Rockies' crest, great geological fault zones splinter the broad valleys. Most prominent is the Rocky Mountain Trench, a fissure from 2 to 10 miles (3 to 16 km) wide in the Earth's crust which separates the Rockies from the older western ranges in British Columbia. The massive proportions of this trench, 800 miles (1,300 km) long in Canada alone, and of the Purcell Trench, which joins it 200 miles (320 km) north of the Canadian border, seem to create a three-dimensional mirror of the mountains of Banff and Jasper, as deep as the cliffs are high.

At the crest, summit shales more than 12,000 feet (3,700 m) above sea

level are stacked like oversized sedimentary pancakes. Their striated massifs and undulating cliffs tilt hazardously from the vertical, while their horizontal surfaces run for miles, as if they could link the ancient past to the unknown future.

The Rocky Mountains stretch more than 3,000 miles (5,000 km), extending northward from central New Mexico through Canada and across Alaska. They reach their highest point – 12,972 feet (3,954 m) – on Canada's Mount Robson, 48 miles (77 km) northwest of Jasper. For most of their length, the Rockies define the Continental Divide, the great watershed separating the rivers that drain into the Pacific Ocean from those that drain into the Atlantic and Arctic oceans.

At the heart of the Canadian Rockies, in the middle of the Banff and Jasper parks, lies the immense Columbia Icefield, a remnant of the great ice shield that blanketed most of Canada 10,000 years ago. Not a single glacier but an astonishing assembly of them, Columbia's 100 sq. miles (250 km²) of icefield smother the mountains and dig deep into their

valleys with tentacles of alpine glaciers. Up to 2,500 feet (750 m) thick at some points, the icefield is buried under about 200 inches (500 cm) of snow every year. Its featureless white surface is gashed by gaping crevasses. Glacial meltwater at the foot of such crevasses is kept from freezing only by its own rushing motion.

Forged by a process that echoes the formation of sedimentary rock, glacier ice is composed of snow crystals that compress under their own cumulative weight and harden. The tremendous pressure of the glacier's weight against the steely bedrock of the mountains creates friction, which heats the lower surface of the glacier and melts it just enough to lubricate the entire glacier's forward motion.

From the Snow Dome of the Columbia Icefield, endless meltwater flows in all directions: via the Columbia River to the Pacific Ocean, via the North Saskatchewan River to Hudson Bay and the Atlantic Ocean, and via the Athabasca, Slave and MacKenzie rivers to the Arctic Ocean.

Jasper's Athabasca glacier is moving in two ways at once. It is

When warmer climates force a glacier into retreat, a natural dam results from the debris at the glacier's head. Copious meltwaters collect and can form lakes such as Moraine in Banff (left).

THE COLUMBIA ICEFIELD

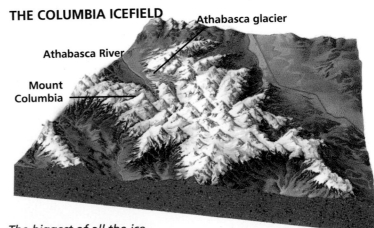

Athabasca glacier

Athabasca River

Mount Columbia

The biggest of all the ice caps in the Rocky Mountains is the Columbia Icefield in the Banff and Jasper parks. Its ice mass reaches to a depth of 1,000 feet (300 m) in places and feeds so many rivers that it is known as the hydrological apex of North America.

receding gradually because it is melting faster than the accumulation of new snow at its head can replenish it. However, since the glacier is still 1,000 feet (300 m) thick in places and fed by substantial annual snowfall, its retreat is extremely slow.

Simultaneously, Athabasca glacier is sliding down the mountain, and, for the first few miles of its flow, the meltwater may not be water at all. Glaciers flow as frozen rivers, similar to fields of molten lava. As Athabasca's ice flows over cliffs, rocks as big as railroad cars split from its face and crash to its base – if these frozen lumps were to split off a glacier and tumble into tidal waters, they would be called icebergs.

Not everything moves at the lumbering pace of glacial ice – a yard or so a day. An avalanche of snow can streak downhill at 100 mph (160 km/h), pushing and compressing the air in front of it into tornado-force blasts that snap mature trees like matchsticks. Rocks also respond to the force of gravity. Blocks of soft sedimentary rock routinely weather and break loose, fanning out from the base of the mountain in vast fields of debris called scree or talus.

The mountainous wilds of Banff and Jasper are a haven for rare bighorn sheep. Once common from New Mexico to British Columbia, bighorn now dwell only in isolated pockets of wilderness, to which they have adapted by developing spongy footpads that give them a firm foothold on bare rock. While they summer in the highlands, harsh winter conditions force them down to the foothills, where domestic stock now graze. Safely hidden from predators among the rocky crags, bighorn sheep easily fall victim to the diseases of domestic livestock, and overcrowding also threatens them with disease.

During the mating season, male rams, weighing up to 250 pounds

"Rocks as big as railroad cars split from the face of the Athabasca glacier and crash at its base"

(110 kg), vie for breeding rights by butting heads. The horn-splitting crashes can be heard half a mile away, and the stately combatants often turn from their confrontations with bloody noses and rolling eyes. But nature has bestowed some protective safeguards on the bighorn sheep. To withstand the repeated head to head clashes, the sheep's skull is porous and double-layered, and thick facial hide absorbs and cushions the blows.

Banff and Jasper also harbor the mountain goat, whose smaller, spiky horns and whiter, long-haired coat distinguish it from the bighorn. Both animals live off the meager mat of alpine vegetation that is briefly exposed by summer's snowmelt – a mixture of dwarf plants, mosses and ground cover that have adapted to the thin soil, brief growing season and harsh surroundings. Certain high-mountain plants, such as alpine buckwheat, grow masses of tiny hairs on their leaves and stems; the hairs block or slow the passage of the wind over the plant and thereby restrict its loss of moisture.

Life throughout these harsh wildlands is arranged in layers that are determined by the growth of vegetation. On the highest meadows live mammals such as marmots and

pikas, or conies, grazing among the grassy pockets of scree and talus. Smaller pikas, sometimes nicknamed rock rabbits, gather grasses and sedges and make little caches beneath rocks, where in winter they eat their carefully stashed hay beneath an insulating blanket of snow.

The next mountain layer belongs to the sheep and goats, and not far below them are the elk, properly called wapiti. Most regal of the large mammals here, these members of the deer family are second in size only to the moose and are distinguished from other deer by their size, yellow rump and the dark mane around their neck. Elk weigh more than 800 pounds (350 kg), and a male's antlers can have a spread of more than 5 feet (1.5 m).

Bulls that grazed the alpine slopes together peacefully through the summer become fierce adversaries when they descend to the valleys in the fall and begin to assemble their harems. For weeks, they parry and shove, bugling majestically and eventually lowering their horns and charging rivals at full speed. Repeatedly victorious males are reduced to exhaustion by the mating season and rendered uncharacteristically vulnerable to bears. Sometimes the bulls' antlers lock inextricably during combat, and both beasts starve to death.

No more sustained stretch of stunning alpine scenery graces the North American continent than that of Banff and Jasper, where towering snowy peaks are reflected in forest-rimmed alpine waters. Lake Louise in Banff displays an almost magical serenity. Draped in glaciers, Mount Victoria rises solemnly behind it, forming a somber backdrop; from certain vantage points, the mountain and its icy cloak appear to rest delicately on the lake's surface.

But this is not gentle country. The rugged nature of these mountain-encrusted parks has helped them remain truly wild lands.

A sure-footed mountain goat (below) wanders the craggy slopes of Banff and Jasper in search of plants to graze on. The goat's straight spiky horns distinguish it from the curled horns of the bighorn and Dall's mountain sheep.

Ever watchful for predators such as eagles, hoary marmots have a warning cry that has earned them the name of "whistle pigs." In spring and summer, marmots feed voraciously on mountain plants to build up layers of fat to last them through the long winter hibernation.

MOUNTAIN LIONS OF BANFF & JASPER

The biggest members of the cat family in North America, mountain lions survive in habitats as diverse as the snowy parks of Banff and Jasper, the parched deserts of the southwestern United States, and the humid subtropical conditions in Florida. Female mountain lions measure up to 8 feet (2.4 m) long, including tail, and males are up to 9 feet (2.7 m).

Mountain lions are stealthy solitary animals which hunt mostly at night and are seldom seen. Deer are their main prey, and one large kill can last a lion several days. In Banff and Jasper, the lions also hunt hares, beavers, mountain goats and sheep. Unlike other big cats, mountain lions do not roar. They utter long and unearthly howls which echo around the remote wildernesses in which they live.

Female mountain lions give birth to one to six young, usually two, in a sheltered den. They care for their cubs alone and teach them how to stalk prey and become powerful killers.

The mountain lion can run fast, but only over short distances. It generally stalks its prey, then makes a final dash to pounce on its victim, killing it with a bite to the neck. Mountain lions are also good climbers and may take cover in a tree to watch for prey.

A snowshoe hare falls victim to a mountain lion (left). Like its smaller cousin the lynx, the mountain lion hunts year round through the rugged winters of Banff and Jasper.

Cougar, puma, *panther and painter are just some of the names by which the mountain lion is known over its range – the widest of any mammal in the Americas.*

Wood Buffalo National Park

"A wetland wilderness of astonishing variety"

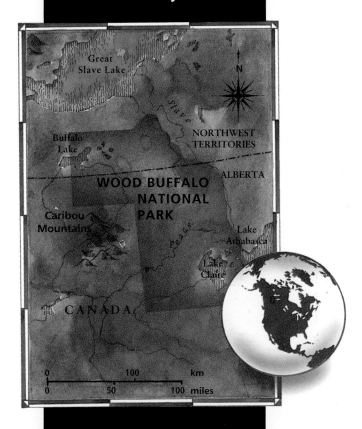

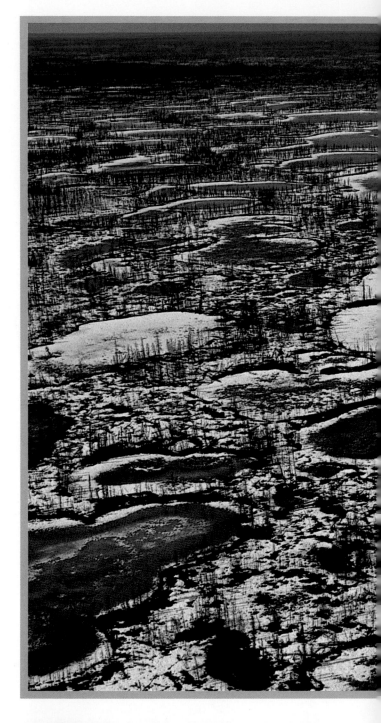

Wood Buffalo National Park is the living definition of big. Originally set aside as parkland to protect the wood buffalo from extinction, it is so spacious that its herd of five to six thousand bison, the largest free-roaming bison herd in the world, can virtually disappear in its vast plains. Larger than all of Switzerland, eight times bigger than

A vast expanse of land with abundant surface water makes the wildlands of Wood Buffalo a natural haven for wildlife.

Yellowstone National Park in Wyoming, Wood Buffalo is Canada's largest park and the second largest in the world. It covers 17,400 sq. miles (45,000 km²), straddling the boundary between the provinces of Alberta and the Northwest Territories south of Great Slave Lake.

Three separate and magnificent wild environments could be created from Wood Buffalo's parklands.

Its rugged uplands feature sprawling, fire-scarred forests of spruce, pine and aspen, characteristic of extreme northern latitudes. A fertile plateau, pocked with bogs and embroidered with meandering streams, unrolls at the lower elevations. Finally, the Peace-Athabasca Delta, one of the world's largest freshwater deltas, encloses shallow lakes, marshes and the continent's largest grass-and-

sedge meadows in a wetland wilderness of astonishing variety. Situated in the park's southeastern corner, the immense delta abounds with waterfowl in spring and summer. Converging here from four North American flyways, the swans, ducks and geese seasonally number more than a million birds.

Much of Wood Buffalo rests on a bedrock of limestone and gypsum,

46

water-soluble rock that dissolves in seeping rain and groundwater, leaving the underground terrain perforated with sinkholes, sunken valleys and complex networks of caverns and subterranean rivers.

In brilliant contrast to the underground caves' mysterious darkness are the dazzling salt-encrusted white mud flats and saline meadows in the southeastern section of the park – the only such landscape in Canada. As the salty water emerges from underground springs and flows across the plains, it evaporates, leaving behind salt and other dissolved minerals on the land. During dry years, salt mounds nearly 6 feet (2 m) high form around salt springs. Salt-tolerant plants grow in these saline meadows. While most plants die in salty environments, several species here excrete the excess salt they accumulate instead of succumbing to it.

Summer is brief in Wood Buffalo, and both plants and animals rush to complete their reproductive cycles before winter returns. The burst of plant life provides food for the clouds of insects which in turn feed migratory birds – and pester bison, moose and other mammals. Dark-furred moose, with branching

UNDERGROUND CAVES AND CAVERNS

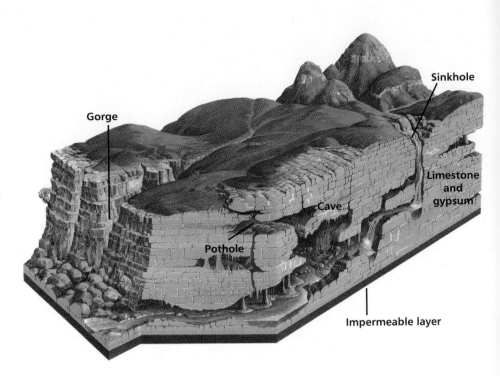

antlers that span almost 6 feet (2 m), thrive in Wood Buffalo's watery lands, browsing on waterside grasses as well as underwater plants in ponds and lakes. Moose represent a large proportion of the diet of timber wolves. These powerful predators hunt moose in packs, generally testing several animals before deciding which one to pursue.

In addition to the moose, the lush grasses of the Peace-Athabasca Delta support most of the park's bison, along with numerous muskrats and beavers. The industrious muskrats take the watery environment as they find it, but beavers alter nature by damming the waterways to create their own safe havens in which to store their food. These large rodents build lodges of mud, sticks and logs which are impenetrable to wolves.

If this varied magnificence were not enough to make this wild place

Wood buffalo are the larger woodland-living cousins of plains buffalo. Although they weigh up to 2,000 pounds (900 kg), these creatures can disappear like magic among the trees.

Black bears spend much of the winter in dens hidden beneath trees or in hollow logs. In spring they emerge to search for food such as berries and eggs as well as small mammals and insects.

The karst landscape in Wood Buffalo is considered the finest example of this kind of terrain in North America (left). The underground structures characteristic of karst are created as water seeps down from above, gradually sculpting the gypsum and limestone into caves, caverns and potholes, sometimes with waterfalls and rivers running through them. Some caves grow so large that they collapse, creating depressions known as sinkholes.

Rare whooping cranes nest and lay *their eggs far from civilization in the wild swamplands of Wood Buffalo (right). There are few such undisturbed wetlands left in North America – a fact that has contributed to the decline in numbers of this graceful bird.*

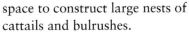

exceptional, Wood Buffalo also contains the only natural nesting ground of the endangered whooping crane. In the 1940s, the entire population of this species dwindled to a mere 17 birds; today, roughly 140 whooping cranes live in the wild, and another 50 or so birds are protected in captivity.

The cranes are white with black-tipped wings, stand 5 feet (1.5 m) tall and typically mate for life. Each year they migrate more than 2,500 miles (4,000 km) from the southwestern United States to breed here. And no smaller habitat would allow them to survive – the territory of each pair of cranes can be anything from 230 to 1,000 acres (93 to 400 ha). In the hospitable muskeg, or level swamp, of Wood Buffalo, they can scatter widely to breed and find ample space to construct large nests of cattails and bulrushes.

In spring the cranes lay two large eggs, which they incubate for 30 days while feeding on marsh plants, frogs and fish. The eggs hatch about two days apart, but the second, younger chick rarely survives. Like many other birds, the parent cranes, accompanied by a single fledgling, leave Wood Buffalo in mid-September to undertake the six- to eight-week flight back to Texas. As the northern winter sets in, bison, ptarmigans and timber wolves are left to reign over this vast domain once again.

Winter is harsh. Temperatures may fall to –40°F (–40°C), and the sun peeps only briefly over the horizen each day. When snow shrouds the landscape, the feathers of ptarmigans, which do not migrate south, turn completely white to lend the bird protective camouflage. Every season, the land and its inhabitants shift and change to accommodate each other, and every year the immense wilderness of Wood Buffalo renews its unique character.

MAZATZAL WILDERNESS

"Inaccessible canyons and sheer-walled cliffs characterize this rugged wilderness"

Hells Hole, Lousy Gulch, Suicide Ridge, Poison Canyon, Hardscrabble Mesa. Only a landscape of the cruelest elements and wildest unyielding nature would generate harsh names such as these, and it is little wonder that they belong to the canyons and gullies of the Mazatzal Wilderness. A vast, empty spread of

A mass of prickly pear cacti typifies the inhospitable nature of this harsh desert land. Yet in the cooler mountains beyond, Douglas firs thrive.

205,000 acres (83,000 ha) in the arid center of Arizona, Mazatzal sprawls eastward from the Verde River, climbs the rugged Mazatzal Mountains, and then plunges to the depths of the Tonto Basin.

The state's largest designated wilderness and one of the largest in the Southwest, Mazatzal's very name reflects its character. Pronounced "madda-zell," the name is either Apache or Paiute and means "the land between" or "the space between." When the original Apache or Paiute residents held up four fingers, representing the mountain peaks, the space between the fingers was termed "maz-at-zark."

The Mazatzal Mountains reach a height of nearly 8,000 feet (2,440 m). Almost all of the southern part of the range is composed of old granitic rock, dating back at least 600 million years to Precambrian times. In the higher eastern portions of the range, outcrops of extremely hard rhyolite and porphyry jasper rocks form steep slopes that resist wear. By contrast, the western side slopes more gently, and at lower elevations the old rocks

are covered by volcanic flows and other volcanic material.

Although Mazatzal is surrounded by semi-arid lowlands, the mountains create islands of more temperate life. On the uppermost vegetation zones are trees such as ponderosa pines and even Douglas fir – a species usually associated with more northerly climes. The presence of the fir here dates back to the ice ages when its range was forced south. But when the continental ice sheets retreated from this region at the end of the last ice age, only high elevations were cool and moist enough to support northern species such as the Douglas fir. These mountain forests of temperate plants are now, in a sense, stranded in this desert land.

The lowest point of Mazatzal, only 2,100 feet (640 m) above sea level, is at the edge of the Sonoran Desert, simmering in temperatures that can rise above 110°F (43°C) in the summer months. Some summers, the temperature exceeds 100°F (38°C) for as many as 80 days in a row. The Sonoran is one of five North American deserts and the only one that extends to an ocean, meeting the Pacific at Baja California.

Humans must struggle harder than plants to survive the searing heat of Mazatzal. It can take a full day to travel on horseback between one waterhole and the next, and the water may not be potable once it is found. Even the water standing in an animal track may come to look attractive.

With its curved limbs raised to the open sky, the saguaro cactus is the area's most characteristic species, perfectly adapted to its desert existence. The roots of the giant cactus are never farther than 3 inches (7.5 cm) from the desert surface and radiate from the cactus's base almost exactly as far as the main stem is high. Through specially adapted root hairs that grow in response to moisture, the roots can soak up as much as

165 gallons (750 liters) of water during a single downpour, enough to sustain the saguaro for a full year. If it survives lightning, wind and occasional frost at the northernmost limit of its range, a saguaro may live to 175 or even 200 years of age, and a 150-year-old specimen may grow 50 feet (15 m) tall.

The saguaro offers shelter to myriad forms of life, day in and day out. Like an apartment building with many stories, it provides varied ecological niches and nesting habitats in close proximity. Gila woodpeckers and their cousins, the gilded flickers, dig holes for their nests in the saguaro's trunk and larger branches. Well insulated by the plant's thick

A saguaro cactus provides a variety of feeding, nesting and shelter sites to an array of desert wildlife. This nesting screech owl claims a hole that was probably drilled out by a gila woodpecker in search of food.

walls, these holes are much cooler in summer and warmer in winter than the air outside.

The woodpeckers excavate new nest holes each year, and their vacant apartments are quickly occupied by sparrow hawks, American kestrels, cactus wrens, screech owls, elf owls and even honeybees. Red-tailed and Harris hawks construct their bulky exterior nests on this cactus, too.

In bloom after *recent rainfall, colorful owl clover carpets the ground around a spiny cholla cactus.*

The roadrunner *(below) runs rather than flies when chasing prey such as rattlesnakes and lizards. This fast-moving bird can reach speeds of up to 20 mph (30 km/h) and kills its victim with a blow from its sharp beak.*

Mazatzal's larger animals include coyotes, black bears and mountain lions, or cougars. These big cats were once widespread in North America, but three centuries of relentless hunting has brought them to the brink of extinction. The species now benefits from the protection provided in wildernesses such as Mazatzal. Here mountain lions can roam, relatively undisturbed, preying mostly on wild deer.

The numerous coyotes and black bears are omnivorous, with a taste for carrion as well as for victims they have hunted themselves. Coyotes eat plants, insects, rodents, birds and reptiles. Agile black bears devour all of these, plus copious quantities of the berries of the manzanita bush.

Although its wild, forbidding lowlands usually vary from dry to bone-dry in climate, Mazatzal blooms vividly after the spring rains with yellow brittlebush, owl clover, blue lupines, golden poppies and a host of other wildflowers. But nature guarantees nothing in this desert

setting, so some flowering plants have prepared themselves for the unreliability of the rains by coating their seeds with germination inhibitors. Until enough rain falls to wash away this coating and assure moisture for the plant's complete reproduction cycle, the seeds will not germinate. Some lie dormant waiting for rainfall for 70 years or more without losing their fertility.

Ironically, Mazatzal can suffer from too much water. When flash floods occur, the steep mountain gorges collect more rainwater than the floors of dry stream beds, known as washes, can absorb. Tumbled boulders and broken cottonwood trees attest to the freakish power of water, whose unpredictable presence proves the wild, unbridled nature of this rough land.

OKEFENOKEE SWAMP

"A brooding stillness wraps the swamp in mysteries as dense and seductive as its vegetation"

D
ark and tangled, echoing eerily with the bellow of alligators and the cackle of cranes, Okefenokee Swamp is the largest freshwater wetland in the contiguous United States. The swamp itself and its nurturing watershed cover roughly 600 sq. miles (1,500 km²) of southern Georgia and northern Florida. The dark water shines like a polished mirror, reflecting the twisted cypresses and elegant water birds, and a brooding stillness wraps the swamp in mysteries as dense and seductive as its vegetation.

Perhaps because it seems so impenetrable, the swamp has become a wilderness of misnomers, with its title and two of its three major habitats mislabeled. More than a swamp, which by definition contains standing water, Okefenokee qualifies as a watershed, since it gives rise to two rivers, the St. Marys and the Suwannee. Its so-called prairies, characteristic watery expanses of vegetation, are actually marshes, and its cypress bays are not bays but dense forests, once logged for their bald cypress trees which usually grow in standing water. The third major habitat, however, is just what its name implies, pinelands of slash and loblolly pine that cannot tolerate standing water.

The present ecological character of Okefenokee was formed 250,000 years ago when the Atlantic Ocean shoreline was 75 miles (120 km) west of today's swamp. An extensive sandbar, eventually about 40 miles (65 km) long, emerged off the coast, and a shallow lagoon formed just

Reflections in Billy's Lake double the dazzling display as low-angled evening sunlight bathes the Okefenokee Swamp. This isolated wilderness is a haven for a variety of wildlife, including huge American alligators.

west of it. After the ocean receded, draining the lagoon, it left behind a shallow sand basin in which Okefenokee Swamp developed. Today's swamp lies about 45 miles (70 km) inland and is shaped like a large, shallow bowl or saucer with two small spots – the outlets of the St. Marys and Suwannee rivers – low on its rim. Its elevation ranges from 105–130 feet (32–40 m) above sea level; by comparison, the Everglades in southern Florida never rise more than 8 feet (2.5 m) above sea level.

Okefenokee is renowned for its peaty brown waters, known as blackwater, with their uncanny, mirroring surface. Over the span of the swamp's existence, blackwater has developed as dying plant life continually drops into the water of the swamp and decays, forming masses of peat. The slowly decomposing peat then releases acids collectively known as tannin, which hangs suspended in the water and gives it a brown cast like that of strong tea, rendering the surface highly reflective.

The submerged layers of peat average from 5 to 10 feet (1.5 to 3 m) in thickness, yet the gases that derive from organic decomposition and become trapped beneath the peat layers sometimes push them to the water's surface and set them afloat. A floating mat of peat may provide a home for airborne seeds that germinate, root and grow on it. Twisting around water-lily roots and those of other aquatic plants, the roots of the new plants anchor the floating peat securely to the bottom and transform the raft into a floating island of decayed vegetation on which even full-size trees may grow.

One meaning given to the name Okefenokee – a local Native American word – is "trembling earth," which refers directly to the floating islands in the swamp that have taken root and become

stationary. The sensation of walking on such a rooted peat island, locally called a "house," has been compared to walking on a water bed.

The other islands that grace Okefenokee are firmly rooted in the distant geological past rather than in botanical decay. Approximately 70 islands are known throughout the swamp, 60 of them large enough to have names. The larger, outlying islands at the edge of the swamp are draped in thick pine forests, while cypress trees cover the small islands deep in Okefenokee's interior.

Many of the larger islands are actually the fragmentary remains of ancient ocean sandbars that took shape during the ice ages before continental ice sheets ground the massive North American rock to

"No creature can be confident that almost imperceptible stirrings in leaves will not explode to reveal the death-dealing jaws of an alligator"

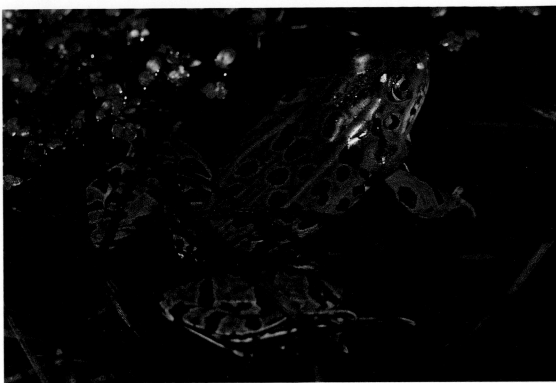

Alligators at rest can look like fallen versions of nearby swamp trees, but these great reptiles can charge with surprising bursts of speed to seize unwary deer at the water's edge. The bellowing roar that often accompanies a charge sounds eerily prehistoric.

Leopard frogs are among the multitudes of amphibians whose croaking calls fill the air in the Okefenokee springtime. They are preyed on by everything from alligators and snakes to stalking herons and ibises.

granular material. As pieces were washed seaward, the relentless tossing of the waves reduced them still further, breaking the grains into sand. Vast quantities of this sand were deposited on what was then the relatively shallow offshore continental shelf. Today, this shelf is the coastal plain in which nestles the gently curved saucer of Okefenokee.

Where sand rather than peat lines the floor of Okefenokee Swamp, clear water sparkles in the dappled light, but the absence of minnows proves that clear waters are not the rule here. In fact, no minnows at all are found in Okefenokee, despite the fact that 9 species live in the river drainages surrounding the swamp, and 69 species have been identified throughout Georgia. Moreover,

though they are regionally abundant, few species of the sunfish family inhabit the swamp. No natural physical barriers prevent the upstream migration of these missing species into Okefenokee, and all species of fish found in the Okefenokee waters are also found in the rivers and streams of the surrounding coastal plain. But the swamp's waters are, on average, significantly more acidic than the surrounding waters, and fish experts have discovered that some of the missing species avoid waters of such acidity.

The 40 or so fish species that do inhabit the swamp include the chain pickerel, largemouth bass, warmouth, gar, mudfish and mosquito fish, and while the range of species may be limited, the numbers of fish are

staggering. As its name suggests, the diminutive mosquito fish has evolved through a direct relationship with the insect whose breeding females bite warm-blooded creatures, including humans, to extract the blood they need for their reproduction process. Mosquito fish feed in turn off the insects' abundant larvae.

Historically, the Florida panther was the major predator of the white-tailed deer that thrive in the swamp. Now this big cat is extremely rare in all of its former range, but there are plans to introduce more animals into Okefenokee and create a protected self-sustaining population there.

Another awesome Okefenokee predator takes the occasional deer but feeds primarily on snakes, fish, turtles and other small swamp

animals. The American alligator is one of Okefenokee's most ancient animals whose lineage goes back 200 million years to the days when scaly creatures, from lizards to dinosaurs, ruled the Earth.

Alligators dominate the swamp's aquatic life today. As opportunistic feeders, they will eat almost any of the swamp's inhabitants, including their own young, and forage onshore as well as in the dark waters. No stalking shorebirds – herons, egrets, ibises – or swimming muskrats can ever be confident that the almost imperceptible stirrings in nearby leaves will not suddenly explode to reveal the death-dealing jaws of a predatory alligator.

Although a mere 6 inches (15 cm) long when hatched, alligators can grow to be 12 feet (3.5 m) and weigh more than 650 pounds (300 kg). These seemingly slow and clumsy creatures are normally observed basking in the sunshine, but they can move with extreme speed and attack with great power. When subduing their prey and in the males' courtship battles, they roll rapidly in the water, turning their bodies into mighty levers to augment the vicelike grip of their elongated jaws.

Another group of predators reinforces the swamp's aura of mysterious wildness in a different manner entirely. Three carnivorous plants – pitcher plants, bladderworts and sundews – sustain themselves in part by eating insects, a predilection determined by their habitat. Meat eating enables them to grow in soil

Perched motionless with wings partly spread, this anhinga is drying out its feathers; they lack the water-repellent oils that keep the feathers of most water birds from getting soaked. Known locally as water turkeys, anhingas evolved from woodland-dwelling ancestors.

FLOATING ISLANDS

Large-leaved spatterdock lilies cover large areas of Okefenokee's water surface (left). They grow on the mats of peat that float on the water and give the Okefenokee Swamp its special character. The peat mats form from deposits of plant matter that settle on the swamp bottom. Gases emitted during the decaying process can force the peat upward and create floating islands where plants, and even trees, can take root (below). The average thickness of the peat islands is about 5 feet (1.5 m), but in places they are as much as 15 feet (4.5 m) deep.

that is too low in nitrogen to sustain plant life. In order to gather the nitrogen they need to survive, pitcher plants and sundews use attractant fluids, and the bladderwort employs entrapping air sacs to lure hapless insects into its grasp. Pitcher plants resemble straightened cornucopias, but function in reverse. Stiff hairs lining the cone allow insects to travel down it, but not up again; eventually they drown in the pitcher's pool of fluids, and their soft body parts are absorbed into the plant's tissues.

At present, only 15 percent of Okefenokee is composed of the open, watery prairies that define its swamp-like character. The geological saucer is slowly and inexorably filling with organic detritus, the remains of a vital botanical ecology. It is therefore possible that at some time in the future, Okefenokee Swamp could transform itself into a relatively dry landscape, sliced by creeks and rivers, where the earth no longer trembles. But for the present, its tea-colored waters remain, and provide a unique home for a remarkable collection of North American wetland wildlife.

Floating islands of vegetation

Peat mats

Gases forcing peat upward

MONTEVERDE CLOUD FOREST

"An abundance of life unrivaled in the western hemisphere"

Perched high in the Tilaran Mountains in the north of Costa Rica, the Monteverde Cloud Forest Preserve is a luminous wilderness of exquisite orchids, sluggish sloths and exotic parrots. This tiny wilderness covers only about 20 sq. miles (50 km²), but contains an abundance of life

In this Costa Rican cloud forest, where streams and rivers abound, there is an incredible diversity of plant and animal life.

unrivaled in the western hemisphere. Nature itelf seems more fertile in this preserve, which boasts more than 2,500 species of plants including 200 different species of orchids alone.

Costa Rica occupies a narrow volcanic rise separating the Caribbean Sea and the Pacific Ocean at the midpoint between North and South America. It covers an area of about 19,600 sq. miles (50,700 km²) and includes two distinct coastal environments as well as watered mountain forests such as Monteverde.

As warm, wet air masses waft westward off the Caribbean Sea, the Tilaran Mountains force them to rise and cool. As a result, the air masses lose their capacity to hold and transport moisture, and they release it as precipitation which nourishes the forests below. Since the Caribbean air masses have lost most of their moisture by the time they reach the crest of the mountains, the western, Pacific, slopes of the Monteverde area, and of Costa Rica generally, are not as wet and verdant as their eastern counterparts. These western slopes become progressively drier as they near sea level, and their lower reaches support tropical dry forests; lower still, dry chaparral similar to that of southern California prevails.

Within the Monteverde preserve, the dense forest canopy all but blocks direct sunlight from the forest floor. So lush is the canopy vegetation that plants even grow on other plants.

Bromeliads, for example, thrive high in the canopy on the branches of trees. These members of the pineapple family create their own microhabitat in the forest, trapping and holding rainwater in their rigid, upthrust foliage. Like a hanging reservoir 200 feet (60 m) above the forest floor, a single bromeliad plant can become the focus of life for bats, salamanders, tree frogs, katydids, beetles, snails and spiders, as well as the water-dependent larvae of insects such as gnats and mosquitoes.

More than 100 types of mammals and 400 bird species inhabit the tiny Monteverde preserve. The mammals form a varied menagerie of wild cats such as jaguars and ocelots, spider and black howler monkeys, three-toed sloths, banded anteaters and collared peccaries, which belong to the pig family. The birds include a flurry of brilliantly colored species of hummingbird, some of which are, without their tails, only the size of a human thumbnail.

Iridescent emerald and scarlet, the dazzling quetzal, Costa Rica's most spectacular bird, is endemic to the mountain forests of Central America and grows to about 12 inches (30 cm) in length. Overlying the main tail, long feathers called coverts provide an impressive train measuring some 35 inches (90 cm). At the end of the breeding season, male quetzals shed these long tail coverts the way some mammals shed their antlers, reducing their overall length considerably.

Tree-dwelling and solitary, the quetzal lives in the lower forest layers. Despite its striking color and spectacular train, it can easily hide itself in the forest, but its alert response to the loud calls of other quetzals, even to a tape recording of

Cloud forest is a type of rainforest growing at high elevations. Because of the height and the constantly damp air, such forest is often swathed in clouds of mist and fog (top).

These jewel-like golden toads live in a small part of the Monteverde forest. Only the male toads are golden – their brilliant hue attracts the dark green or black females.

Turkey vulture

Swainson's hawk

Migratory route

CENTRAL AMERICA

BIRD FLYWAYS
Areas such as Monteverde are vital stopping-off points for birds migrating from North to South America. Many birds depend on strong thermal air currents over land and are forced to travel over the narrow strip of Central America.

the clucking, reveals its presence to human and animal predators.

Among the largest South American parrots, scarlet macaws can be 33 inches (85 cm) long. These parrots are most often seen as pairs, but they do flock together in groups of 20 or so when moving from nighttime roosts to feeding sites. As the flock moves, pairs fly nearly wing to wing. Surprisingly little is known about the domestic habits of these macaws. They feed in the trees in near silence, eating nuts, fruits, berries and seeds and emitting their raucous, squeaking alarms only when they are disturbed and take to the air.

In complete contrast to the scarlet macaw, the three-toed sloth lives a life of near invisibility, conveniently clothed in a mottled gray fur coat that perfectly matches the bark of its favorite tree, the *Cecropia*. This sloth is so well adapted to life in the trees that its hair grows in the opposite direction from that of most mammals. The fur points downward for the efficient shedding of water when the sloth is hanging upside-down on its hooklike claws. The algae that live in the grooved hairs of the fur, tingeing it green, contribute to the sloth's effective camouflage.

In Monteverde's lush landscapes and constant climate, many animals and plants have evolved together, in a process called coevolution. The perfect match of the sloth's fur and the bark of the *Cecropia* tree is just one example, but nature's infinite sophistication can be seen everywhere in this ecological complex of cloud, forest, water and wildland.

With its showy plumage and long streamerlike tail coverts, the male resplendent quetzal richly deserves its name. Quetzals live in the cloud forest trees, where they search for fruit and insects to eat and make their nests in rotting trunks.

MORNE TROIS PITONS

"A mountainous and little-known land where dense vegetation seems to shut out the rest of the world"

Aparadise of verdant trees, brilliant birds and steaming lakes rising out of the Caribbean Sea, the island of Dominica houses the largest stand of island tropical rainforest in the Caribbean. More than 40 percent forested and the least developed of the five Windward Islands (the others are St. Lucia, St. Vincent, Grenada and Martinique), Dominica is a mountainous and little-known land.

Time seems to have stopped here. The island has no sandy beaches to lure the invasive tourist and no port to draw large ships, so much of Dominica remains unexplored except by the native Carib Indians. In its Morne Trois Pitons National Park, the dense vegetation seems to shut out the rest of the world.

Dominica's craggy cliffs are crowned by Morne Diablotin, which stands 4,774 feet (1,455 m) above sea level and marks the highest point in the Windward Islands. The rainforests of Morne Trois Pitons drift up to blue-green peaks that explain the island's nickname "Switzerland of the Caribbean"; at the highest elevations dwarfed forests, shaped by the wind, supplant the rainforest.

The mountainous interior of Morne Trois Pitons bursts with natural surprises. Giant ferns line the banks of the rushing rivers, and orchid plants grow on the trees, their bright colors glowing in the dim, nearly green light of the canopy. Among the most stately trees of the Morne Trois Pitons forests are the gommier or gumwood trees, named

Rainforests thrive in Dominica's Morne Trois Pitons National Park. Volcanic mountains force the trade winds to drop their moisture as rainfall, assuring a plentiful supply for the dense plant growth. Hidden in the forest are many waterfalls and deep clear pools – this one is known as the Emerald Pool.

for their aromatic gum, which grow to more than 100 feet (30 m) in height. Island people still fashion their traditional dugout canoes from this majestic species, and the national bird of Dominica, the endangered imperial parrot or sisserou, nests in holes high in the trunks.

Native to the island, the sisserou does not live naturally in the wild anywhere else in the world. It is one of the largest parrots, measuring up to 20 inches (50 cm) long with a

fabulous 30-inch (75-cm) wingspan. The parrot's gorgeously colored wings are green and red, its tail and back a brilliant dark green; red eyes gleam in its iridescent head. A broad, deep violet band crosses the back of its neck, and its breast is bright purple. Sisserous feed while hanging head down – a posture well suited for plucking the fruit of the tall kaklin trees they seek out.

The wild population of sisserous now numbers only 50 or 60, and their continued survival is at risk. When Hurricane David ravaged Dominica and its forests in 1979, its 150 mph (240 km/h) winds threatened many native species with extinction. It levelled many of the gommier trees in which sisserous were nesting and blew the entire year's crop of nuts and fruit off many of the island's other trees. The parrots that escaped the lethal winds were driven down into the lowlands in search of food. Many were illegally

captured and offered for sale on the black market as cage birds.

For the sisserous that currently haunt Dominica's higher mountain reaches, volcanic eruption looms as a permanent long-term threat. Dominica has never outgrown its volcanic origins, and the volcanic forces lurking below the island bubble to the surface perpetually in Morne Trois Pitons, particularly in the barren Valley of Desolation.

Noxious underground gases have turned its bare rocks every imaginable shade of red, yellow and brown, and minerals that have leached to the surface color the valley's waters both jet black and cloudy white. Volcanic ground vents give forth deafening roars, and a dramatic waterfall plunges over one end of the nearby Titou Gorge, a volcanic fissure that has filled with water.

Often hidden by its self-generating veil of water vapor, Boiling Lake in the Valley of Desolation is the world's

The aptly named Boiling Lake of Morne Trois Pitons is a flooded fumarole some 260 feet (80 m) wide in the bleak Valley of Desolation. Billowing clouds of steam rise from the seething waters of this eerie lake, which is kept bubbling by the volcanic heat beneath.

A BUBBLING LAKE

Boiling Lake

Water permeating through porous rock

Fumarole

Heated rock

Magma

In a fumarole such as Morne Trois Pitons' Boiling Lake, groundwater from the surrounding mountain region percolates through porous rock and is heated by the magma below. This super-heated water is then forced up through a vent to the surface as steam.

largest fumarole. Fumaroles occur where groundwater that has been excessively heated underground returns to the Earth's surface as steam, lacking enough moisture to flow as water. Where acid gases transform solid rock into clay, as at Boiling Lake, fumaroles bubble in wet mud; smaller, viscous fumaroles are known as mudpots.

Magma, a molten material which lies unusually close to the surface of Morne Trois Pitons, heats the subterranean groundwater deep in the Earth, where it is subjected to enormous pressure. As the steam pushes up toward the surface, however, the pressure on it slackens, and it expands with tremendous force, literally bringing Boiling Lake to a boil.

The abundant rainfall in Dominica is inextricably tied to its mountains, as the disparity between the rain on the coast and in the mountainous interior proves. On its windward coast, Dominica registers 70 inches (1,780 mm) of rainfall per year; the high mountains receive more than 236 inches (6,000 mm). This phenomenon is called orographic rainfall. The mountains force the moisture-laden air to rise and cool; as it does so, it releases its moisture as rain.

The liberal orographic rain that pours onto Morne Trois Pitons and other mountainous areas nourishes the complex ecology of this luxuriant island. Its natural riches are protected by the very wildness of the rugged, densely forested terrain that, so far, has kept the world at bay.

Roots of a giant fig stream earthward like a forest of mini tree trunks. In fact, this fig cannot support itself structurally, but depends on the tree it parasitizes. It may eventually strangle the host tree, compressing its trunk until its cells can no longer transport water and nutrients.

"Dominica has never outgrown its volcanic origins, and the volcanic forces lurking below the island bubble to the surface in Morne Trois Pitons"

THE
TEPUIS OF
VENEZUELA

"Mysterious flat-topped mountains rise from this ancient wilderness"

The only way to penetrate the remote tropical forests of southern Venezuela is by river. About 400 miles (650 km) due south of Caracas on a tributary of the upper Orinoco, little can be seen from the water beyond the continuous green wall of the trees. But at a bend in the river, a huge rock suddenly looms into view. Blue in the

Tepui

Roraima group of tepuis

Guyana Shield

0   100   200   300   km

0   100   200   miles

CARIBBEAN SEA

VENEZUELA

Orinoco

N

COLOMBIA

Tanaima

Tepui

Mount Roraima

GUYANA

BRAZIL

distance – it is some 30 miles (50 km) away – its sides climb sheer from the surrounding forest in one enormous step to a broad, level summit that is suspended hundreds of yards in the air. The extraordinary rock is Autana, and it is one of more than a hundred mysterious, flat-topped mountains that rise in this ancient wilderness region of South America.

Often described as rock towers, mesas or table mountains, they are best known by their local Penom Indian name "tepuis." The greatest concentration of them lies in and around the Canaima National Park of southeastern Venezuela, but others are scattered across the rest of southern Venezuela, the northern frontier of Brazil, Guyana and Surinam. All have

Enveloped in cloud, the rim of Auyan-tepui falls sheer to the forest far below. Auyan-tepui is the largest in area of scores of similar table mountains that rise like dislocated slabs of earth in the region south of the Orinoco River. The isolation and mystery of their broad summits have long inspired legend and adventure.

steep, perpendicular sides, rising as much as 5,000 feet (1,500 m) above the land beneath, sometimes in a single slope, sometimes in two or three giant steps. Their flat summits form elevated portions of land almost as isolated from one another and from the ground far below as islands in an ocean. Remote, awesome, often shrouded from view by swirling mist and cloud, many of these lonely summits have yet to be explored.

Autana is an outlying tepui, a long way from the nearest flat-topped peak. But from high viewpoints in Canaima, the shapes of tepuis of various sizes stud the horizon. The giant slab of Mount Roraima, at 9,220 feet (2,810 m) the highest of them all, commands the view to the east. Other much smaller tepuis, separated by miles of intervening land, stand like blunted pinnacles. It is as though an enormous scythe has been swept over the land, slicing away the tops from a range of scattered mountains.

The explanation for this dramatic landscape lies in its geological history. The rock of the tepuis is mainly sandstone with intrusions of crystalline material such as quartzite. Geologists believe the sandstone formed at least 1.8 billion years ago, making it one of the oldest sandstones in the world. (It directly overlies the so-called Guyana Shield, the oldest rock in all South America.) The layers of sandstone laid down by water and wind accumulated to a depth of up to 8,200 feet (2,500 m). Much later, these layers came to form a vast plateau surrounded by lower land.

When the plates of the Earth's crust carrying South America and Africa started drifting apart about 135 million years ago, the plateau was probably wrenched and fractured by these forces. Some blocks may have been lifted upward, while rivers crossing the plateau and tumbling down its sides cut deep valleys

between them, especially along the weakened fissures. As the climate became wetter, so erosion intensified. In time, the plateau became ever more dissected until all that remained were the isolated blocks of today.

Millions of years from now they, too, will have been eroded out of existence. Rainfall, already high in this region of South America, is often torrential on the tepuis. Air rising over the summits cools and sheds up

> "Remote, awesome, often shrouded from view by swirling mist and cloud, many of these summits have yet to be explored"

to 157 inches (4,000 mm) of rain every year, in cloudy drizzle as well as violent electrical storms. Streams etch deeply into the terrain and at the summit edge plummet over the rock wall on all sides. These numerous waterfalls carve backward into the tepuis as they cascade down to the lowland, and the sides of the rock are eventually eaten away.

From the surrounding lowlands, people have long viewed the flat

heights of the tepuis as worlds set apart from their own. Native legends have grown up concerning many of these cloud-shrouded peaks, and some, like Autana, are considered sacred. A remarkable cave runs straight through Autana, opening high in the rock wall on each side – a natural wonder said to be the home of a monster.

Nobody has yet found any monsters, but the isolation of the tepuis is strikingly reflected in their known flora and fauna. Around half of all the plant species that occur on them are endemic – they do not exist anywhere else in the world. They include hundreds of species of orchids, bromeliads and sundews, as well as shrubs and low trees. Many are confined to a single tepui.

Animals unique to the tepuis have also been discovered, including endemic insects, amphibians such as the minute toad *Oreophrynella quelchii*, and birds like the greater flower-piercer and the tepui spinetail. Because they are adapted to survive in the summit conditions, including a climate of cold nights and warm days, it is difficult for many tepui plants and animals to spread farther afield even if they could escape their highland fortress. For beyond the cliffs lie tropical lowlands with a dramatically different climate.

Even birds of the tepui summits would be unlikely to fly far from their home region. Yet there are birds living high on the tepuis with close relatives at least 435 miles (700 km) away in the Andes Mountains. The

Drenching rain on Auyan-tepui feeds Angel Falls, the highest waterfall in the world. In two leaps, the water plummets 3,212 feet (979 m) over the edge of the tepui, from the coolness of high altitude to tropical heat at the foot. The total drop is nine times that of Africa's Victoria Falls.

Paramo seedeater and the white-throated tyrannulet, for example, are widespread in the Andes, and both have distinct races out on the tepuis. But, as far as is known, no highland ever linked the tepui region with the Andes, so how could these normally sedentary birds have made their way across such a vast environmental barrier as the tropical plains?

The answer may be that, some time in the past, the barrier was not there. Changes in climate brought temperate conditions to the lowlands, and the birds naturally advanced across them. When the climate warmed again, the lowlands became unsuitable once more. Those populations living near tepuis retreated upslope to find the right conditions. Some reached the mountain tops and managed to survive, but in isolation.

Many tepuis rise abruptly from jungle-cloaked land. The forests continue onto the lower slopes, coating the wedges of weathered debris that skirt the cliffs. Even the cliffs themselves may be laced with vegetation clinging to crevices and ledges. The walls of Autana are thick with plant life in places and noted for the abundant tarantulas that patrol the vertical rock face in search of prey. As the cliffs rise ever higher, so the plants that cling to them need to be able to tolerate the cooler temperatures of altitude as well as a combination of murky daylight for much of the time and intense sunshine when the mists clear.

On most tepuis, the vegetation of the elevated summits looks quite different from that toward the bottom. Mount Duida which rises from the rainforests southeast of Autana is topped with low bushes, swamp patches and hardy cushion plants – a distinct contrast to the towering trees at its foot.

Dense, low vegetation also clothes much of Auyan-tepui, which lies within Canaima National Park. With

FORCES OF EROSION

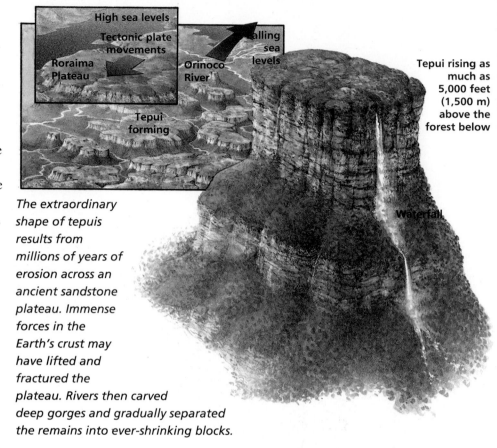

The extraordinary shape of tepuis results from millions of years of erosion across an ancient sandstone plateau. Immense forces in the Earth's crust may have lifted and fractured the plateau. Rivers then carved deep gorges and gradually separated the remains into ever-shrinking blocks.

Tepui rising as much as 5,000 feet (1,500 m) above the forest below

a surface area of about 270 sq. miles (700 km²), Auyan-tepui is the most extensive of all tepuis. Streams racing across its broad summit have cut deep gorges into its heart, and one of them tumbles over its rim to form Angel Falls, the highest waterfall anywhere in the world. By the time the plummeting water has descended from the lofty heights to the tepui's forested base, most of it is a misty veil of spray.

Roraima and its neighbor Cuquenan, though higher, are not nearly so large in area as Auyan-tepui. Nor are they so well vegetated. Little soil exists on their blackened summits, and intense rainfall has leached most of the mineral nutrients from the weathering sandstone. Their surfaces are gnarled and weirdly sculpted, sometimes into almost impenetrable rock labyrinths, with

pockets of plant life lurking mainly in crevices and hollows.

In places, however, clumps of moss and lichen growing on bare rock create miniature platforms on which other small plants can secure an anchorage. Some have even evolved carnivorous habits. Given the paucity of nutrients available to roots, some plants use devices such as sticky traps and slippery pitchers to catch and digest insects instead.

Botanists are gradually learning more about the unique plant life of the isolated summits, but the animals of the tepuis remain little studied. Though a steady trickle of hikers now reaches the top of Roraima every year, great sections of the summit are yet to be penetrated. Meanwhile, some tepuis have still to receive their first human visitors. Who knows what mysteries await them?

On the bleak, rocky surface of Roraima (above), there are few nutrients. Plants have special adaptations to their strange environment.

Plant life on tepuis is radically different from that down below, with many species unknown elsewhere (left). Auyan-tepui is rich in endemic orchids.

MANU BIOSPHERE RESERVE

"Dense virgin forest, with one of the greatest concentrations of wildlife on the planet"

On the eastern slopes of the Peruvian Andes lies a region so remote and so little explored that archaeologists still dream of finding within it long-lost hoards of Inca treasure. But its true riches are of a biological kind. Mantled in dense, virgin forest, it has one of the greatest concentrations of wildlife diversity on the planet. This region of South America is of such global importance that the world body UNESCO declared it should become the Manu

Biosphere Reserve, in recognition of its outstanding ecological wealth.

With an area of over 7,260 sq. miles (18,800 km²), Manu Biosphere Reserve covers part of the Upper Madre de Dios River basin, including all the lands drained by its major tributary, the winding Manu River. The headwaters of the rivers lie in the easternmost wall of the Andes, the Paucartambo Mountains. From the open and windswept slopes around these 13,000-foot (4,000-m) peaks, the streams fall rapidly into rugged

valleys carved into the mountains' flank, filled with ever thicker vegetation and enveloped for much of the time in swirling cloud. These daunting slopes, it is conjectured, could still harbor somewhere the "Paititi" or "lost city" of the Incas. This legendary place is the secret refuge to which Inca people are supposed to have fled, taking with them treasures they saved when the Spanish pillaged their highland realm.

Eventually, the streams cease their turbulent plunge as they reach the

From its winding lowland rivers to its open mountain tops, the Manu reserve on the fringes of the Peruvian Andes is immensely rich in wildlife. It remains one of the most isolated and least explored parts of Amazonia.

foot of the Andes. The rivers become broad and deep and start to meander over flatter ground in the eastern half of the reserve. Their waters are destined to reach the Amazon River, and by now their banks are clothed in

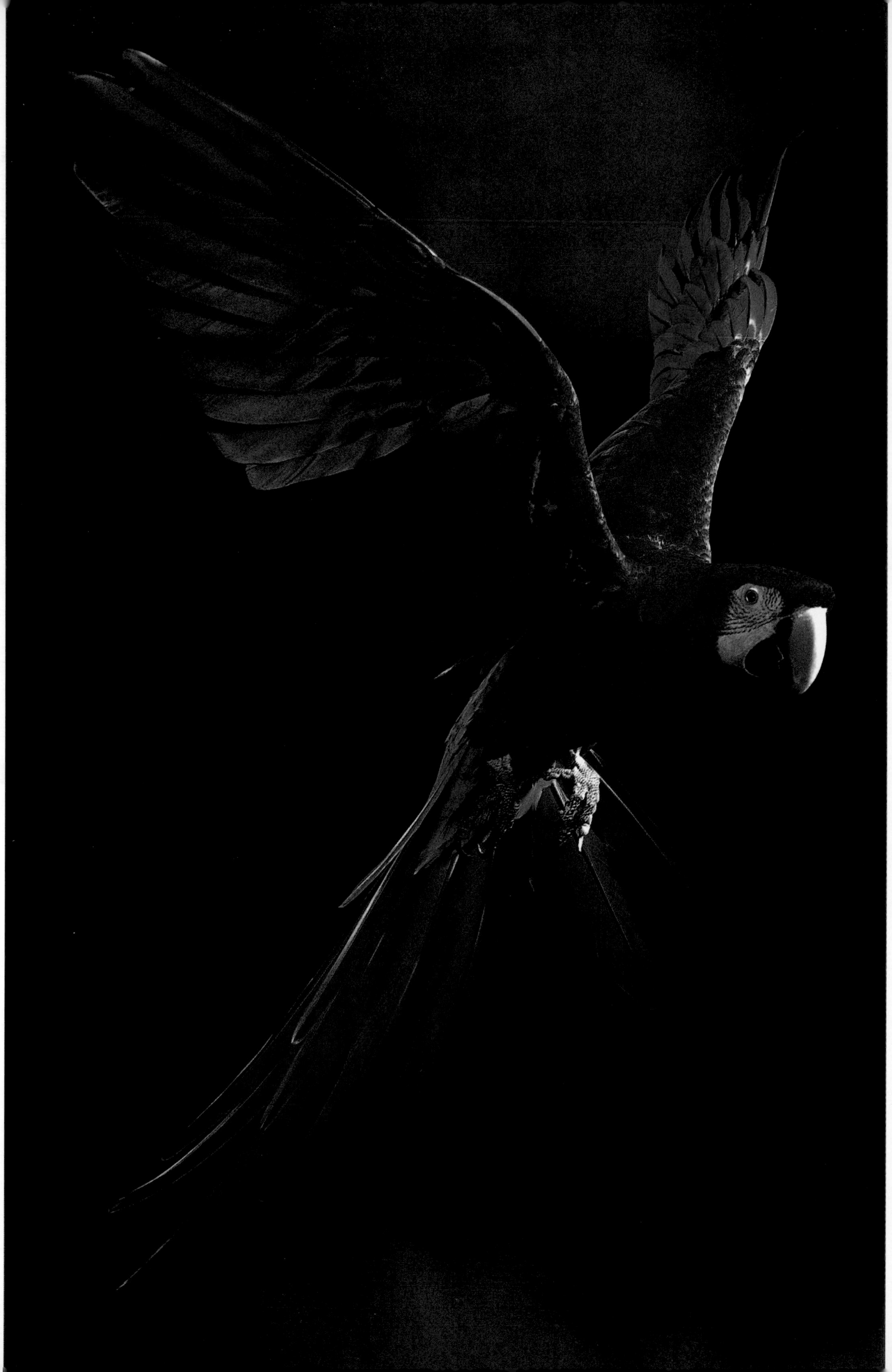

full lowland rainforest, one of the most pristine stretches of continuous forest in all the Amazon region. Few outside people have settled in the area of the reserve, and those scattered tribes for whom the forest has long been home, such as the Machiguenga, Kugapakori and Nahua, still lead for the most part traditional lives, with minimal adverse impact on their natural surroundings.

If one were possible, a roll call of the wildlife species that inhabit Manu would take in a huge proportion of the animals that exist in South America. Inventories of the reserve's fauna, even for larger animals, however, are far from complete. Estimates suggest that there could be 200 mammal species and as many as 90 different kinds of frogs. Bird species found throughout the reserve could number more than 1,000 – an extraordinary 550 birds have been recorded in one area alone. Fishes in the streams, rivers and lakes are highly diverse, and the number of different insects and other invertebrates could easily run to one million or more.

Yet it is not simply diversity that makes Manu so special. Because the forests and mountain slopes have suffered so little human disturbance and exploitation, animals becoming rare in other parts of the Amazon occur in much more healthy numbers in Manu. These include endangered species such as the giant otter, which hunts for fish in the lowland rivers,

and the harpy eagle, which is the reserve's symbol. This raptor is a giant among birds of prey, capable of snatching animals as large as monkeys and sloths from the treetops.

Some of the wildlife ranges widely over the park, from the lowlands up through the lower mountain slopes. This includes foraging and browsing mammals like the agouti and the red brocket deer, as well as hunters like the jaguar and the ocelot. The king vulture, a spectacular scavenging bird that relies largely on smell to detect its food, cruises conspicuously over both the lowland and upland forests. A few highly adaptable animals, among them the puma, occur from the plains right up to the high elevations well above the forest.

But most animals are not so flexible in their habitat preferences. Varying conditions in the reserve associated with different altitudes therefore bring changes in the mix of wildlife.

The flatter, lowland sections below about 2,000 feet (600 m) altitude are hot and humid throughout the year. Rainfall becomes heavier from October to April, but is always high, with frequent downpours from dark thunder clouds that gather over the landscape. Temperatures are around 75°F (24°C), rising to a stifling 97°F (36°C) on the warmest days.

Constant heat and moisture are ideal conditions for luxuriant plant growth, creating the richest of all land

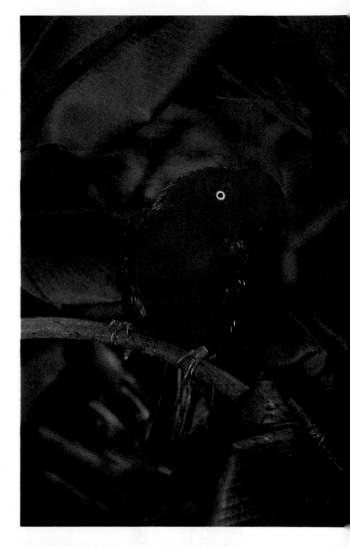

The Andean cock-of-the-rock, with its vivid plumage and curious semicircular crest, dwells in misty, jungle-clad ravines. While females watch, male birds perform dramatic mating displays. They leap into the air, whirring their feathers, flicking their tails and making loud calls.

The spectacular scarlet macaw (left) is one of several members of the parrot family that live in the Manu rainforest. Noisy flocks of macaws gather to feed at fruiting trees.

Colorful fruits attract numerous forest birds, which eat the flesh and later disperse the seeds in their droppings. These fruits have been split open to reveal their seeds (left).

habitats – classic tropical rainforest. Deep in the forest, away from the crowded, sun-soaked "jungle" that typically lines a clearing or river, the air is still and the light is muted. Tall trees rise with columnar trunks before spreading their branches and leaves at the crown to form a continuous green canopy up above. Varieties of cedar, mahogany and kapok mingle with hundreds of other tree species so that it is hard to find two of the same kind standing side by side. The largest have massive buttressed bases up to 10 feet (3 m) across and rise more than 200 feet (60 m) into the air.

Beneath the canopy, twisted lianas dangle from on high, creepers cling to the trunks and branches, and palms spread their fans. Down near the ground, cast in continual shade, thin saplings strive upward, and an untold variety of low-growing plants thrives scattered over the fallen leaves, twigs and other debris.

Though all levels of the forest abound in wildlife, it is the canopy, the forest roof, that harbors the greatest concentration. Here are the bulk of the leaves, fruit, nuts and

> *"Among the hundreds of tree species it is hard to find two of the same kind standing side by side"*

flowers – food for myriad insects and spectacular birds such as cotingas and toucans. Howler monkeys, whose loud calls echo through the forest, roam the canopy feeding largely on foliage. Hordes of plant-eating creatures attract legions of hunters that eat them or prey on one another, including spiders, tree frogs, lizards, boas, woodpeckers, hawks and bats.

Below them deer and tapirs browse on the undergrowth, and large ground birds such as tinamous and currasows scour the forest floor for fallen food.

Moving westward across Manu on to the lower mountain slopes, subtle changes appear in the rainforest habitat. A combination of steepening slopes, thinner soil and slightly lower temperatures makes it increasingly difficult for trees to grow to the heights they reach on the lowland. The canopy becomes less even, more light reaches the forest floor, allowing thicker undergrowth, and different types of flora have a chance to grow.

Eventually, the slopes reach a height where they are enveloped for much of the time in low cloud, bringing even more rain than in the lowlands. Temperatures, still hot by day, become cool at night, when the air is dank. Many of the trees seem twisted and stunted compared with those lower down, and their branches are typically laden with other plants such as lichens, mosses, ferns and beautiful orchids.

This is the cloud forest, home to an especially high diversity of small, colorful birds, including mixed flocks of tanagers that forage through the treetops for berries and insects, and various hummingbirds that sip nectar from the abundant flowers. It is also the stronghold of the vivid red and black Andean cock-of-the-rock, which nests in rocky ravines, and of the spectacled bear, which piles sticks on a tree fork to make its lair.

By an altitude of about 10,000 feet (3,000 m), the forests give way to an open landscape of tussock grasses and shrubs, with reedy pools on the flatter portions. This highland habitat, restricted to the southwestern limits of Manu, is known as "puna." Temperatures are low at this elevation – becoming frigid at night – winds are strong, and there is a marked dry season from May to September. To cope with the high

Solitary, stealthy and formidably strong, the jaguar is the top predator of the South American forests. It hunts mostly at night. Having stalked its prey, which may range in size from a ground bird to a tapir, the jaguar attacks with a mighty leap, aiming to clamp its powerful jaws around the neck of the victim.

mountain conditions, both plants and animals have to be hardy. Thick fur is an asset for small mammals like the mountain viscacha as well as for its enemy, the Andean fox. They share the grassy slopes with a deer that is becoming increasingly rare in the Andes, the Peruvian huemul.

Although the Manu Biosphere Reserve is remote, conservationists are fighting to keep it free from the pressures of logging, gold-mining and oil exploration that are already affecting regions along its borders. The challenge to keep Manu in its natural state means not only maintaining a refuge for endangered species like the huemul and the harpy eagle, but also preserving the entire ecological base of one of the richest forest lands on Earth.

Rivers snaking their way through lowland Manu shift course over the cenuries as they cut new paths and dump sediments transported from the mountains. Abandoned meanders form tranquil oxbow lakes, which make secluded havens for wildlife (below).

THE ZONES OF THE FOREST

Marked changes in vegetation appear with increasing altitude on the mountain slopes in western Manu. The constant heat that bathes the lowland rainforest gives way to cooler temperatures on the middle slopes. Fog and cloud commonly drape the forest here, the canopy of which is lower and less even. Higher still, conditions become too cold for forest trees to develop at all.

Puna

Cloud forest

Increasing altitude
Decreasing temperature

Lowland rainforest

NATIVES OF THE AMAZON BASIN

Since time immemorial, the Amazon Basin has provided food and shelter for people scattered across its rainforests, riverside flood plains and fringing savannas. The many indigenous native tribes include groups such as the Kayapo of the south, the Jivaro of the west and the Yanomami who live in the north of Amazonia. Today, though most of the native cultures have been swamped by often disastrous contact with outsiders, some tribes still lead traditional lives, in close contact with the natural world around them.

Though lifestyles differ from tribe to tribe, most Amazonian people live by a combination of farming, hunting and gathering. Tribal villages are usually surrounded by garden plots cleared by slash-and-burn. When the soil is exhausted, the villagers move to new sites. Various weapons are employed in hunting and fishing, among them harpoons, bows and arrows, traps and snares. Poison-tipped darts are blown with lethal accuracy from long bamboo guns to fell groups of birds and monkeys high in the trees. Other food is simply harvested from forest plants.

A Yanomami Indian prepares a riverside reed for use as a hunting arrow. The forest provides myriad food sources for the Yanomami people. Some, such as birds, monkeys, deer, tapirs and armadillos, have to be hunted. Others, including fruit, nuts, honey, caterpillars and beetle grubs, are simply harvested from the wild and stored in baskets and fiber pouches fashioned from forest plants.

Many Amazon peoples cultivate staple crops suited to the rainy tropical climate such as manioc, plantain and maize. Here, Yanomami women grate manioc to prepare bread for a funerary feast (right).

A Kayapo Indian fishes with a bow and arrow in the calm waters of the great Amazon River (below). Such a method of fishing requires patience, stealth and skill.

ITATIAIA NATIONAL PARK

"The call of the wild is ever present here, where parrots race across the treetops with raucous cries"

BRAZIL

ITATIAIA NATIONAL PARK

Pico das Agulhas Negras

Serra da Mantiqueira

N

0   50   100   km
0   40   miles

ATLANTIC OCEAN

Set among rugged mountains, some 40 miles (65 km) inland from the coast of southeast Brazil, is the Itatiaia National Park. The park is renowned for its luxuriant tropical forest that sweeps up the hillsides, clings to steep valley walls, and crowds around every cascading stream.

Inside the forest, the call of the wild is ever present. In the early morning, when mist still hangs in the air, bands of howler monkeys challenge one

Lush and tropical, Itatiaia National Park encloses a precious fragment of Brazil's once extensive Atlantic Forest. Many of the forest's plants and animals are unknown outside the region.

another across the valleys. Their haunting calls are not so much howls as deep roars like the sound of distant wind. As the sun rises, flocks of parrots and parakeets leave their roosts and race across the treetops with raucous cries. Later, the muted sounds of still noon are broken by the drilling of a woodpecker's beak, and at night the damp air echoes with the peeping calls of tree frogs.

A few centuries ago, sounds like these were commonplace far beyond the boundaries of Itatiaia. Where the bulk of Brazil's population lives today, a huge belt of dense forest used to carpet the plains and mountains abutting the country's eastern coast. Known as the Atlantic Forest, it was a type of ecosystem quite separate from, but no less exuberant than, the Amazon rainforest well to the north.

Charles Darwin was thrilled by the Atlantic Forest in 1832 during his epic voyage on HMS *Beagle*. He wrote: "Here I first saw a Tropical forest in all its sublime grandeur," and remarked upon "not being able to walk one hundred yards without being fairly tied to the spot by some new and wondrous creature."

But already, by Darwin's day, the forest was disappearing. Today all that is left are scattered remnants: precious wildernesses – like Itatiaia – mostly on mountainous land that is too steep for easy cultivation.

The 115 sq. miles (300 km²) of wild country that make up Itatiaia

National Park lie within a large mountain chain, the Serra da Mantiqueira. An offshoot of the range, the Serra do Itatiaia, forms the central highest ground of the park, a cold, open plateau of grassland and bare granite rocks. Rainfall is plentiful for most of the year on the plateau, although it becomes drier in winter, when nighttime temperatures often drop below freezing. Clumps of grasses, thin-leaved bamboo, small shrubs and herbs grow here, providing cover from predators like the savanna fox for small birds, including the rufous-tailed antbird and the Itatiaia spinetail.

Far richer in terms of wildlife are the forests that spread across the bulk of the park to the north and south. At a height of about 6,000 feet (1,800 m), the upper limits of the forest are shrouded for much of the time in cloud. All the forested land of Itatiaia is montane forest, its appearance strongly reflecting the influence of steep slopes and high humidity.

Because of the mountainous nature of the land, the tree canopy is stepped rather than even, and there are more open patches than in a lowland rainforest, allowing pockets of ground vegetation to thrive. Some of the biggest trees, like the massive 130-foot (40-m) jequitiba-rosa, have space to grow horizontal branches and broad spreading crowns. Others

The saffron toucanet perches high in the forest, plucking berries and fruits with its colorful beak.

are prominent not for their size but for golden or violet blooms that, at a distance, coat the trees with vivid color against the green backdrop.

Palms and tree ferns are common in the forest, especially in the damper areas near the mountain streams. Patches of bamboo, however, are more common in drier, sunlit places and form extensive groves up toward the tree line. But the forests are most impressive for the incredible abundance of plants that establish themselves high above the soil, supported on trunks and branches. Plants with such a habit are called epiphytes. Many of these plants,

"An incredible abundance of plants grows high above the soil on tree trunks and branches"

In early morning, and often later in the day, mists and cloud hang over the verdant landscape of Itatiaia. Much of the forest is on rugged slopes above 3,600 feet (1,100 m).

including ferns, mosses and orchids, draw their moisture either from rainwater running down the tree bark or directly from the humid air through delicate dangling roots.

Bromeliads, however, store their own water supplies. The smooth radiating leaves of these large plants pack tightly together at the base to form a container for falling rain. Itatiaia is renowned for its bromeliads, three-quarters of which occur nowhere else but in the remaining fragments of the Atlantic Forest. They crowd at such density on the biggest tree branches that the stout limbs look in danger of collapsing under the weight.

Like many of the plants, a large proportion of the animal inhabitants

of Itatiaia are unique to the Atlantic Forest ecosystem. They include rarely seen animals of the treetops, such as the wooly spider monkey – the largest monkey of the New World – and the leaf-eating maned sloth, as well as bold, conspicuous creatures like the red-breasted toucan, which feeds largely on fruit and nuts.

Other inhabitants are much more widely distributed across the forests of South America, including ground foragers like the paca and the tapir, hunters such as the jaguar and ocelot and powerful raptors like the harpy eagle. As more and more rainforest is destroyed, protected parks like Itatiaia are an increasingly important refuge for such wildlife.

SAVING THE GOLDEN LION TAMARIN

With its vivid coloration and opulent mane, the golden lion tamarin is among the most celebrated, yet most endangered, of the Atlantic Forest's unique animals. It used to thrive in the lowlands near Rio de Janeiro, but today its wild population has dwindled to only 500 individuals.

Strenuous efforts are being made to protect the remnants of the monkeys' habitat. And, with the help of zoos around the world, Brazilian biologists have set up a special reintroduction program. Captive-bred tamarins have already been released into a protected area of the forest in the hope that they can survive and form a stable breeding group in the wild.

THE
BOLIVIAN
ALTIPLANO

"An immense level plain – a frigid desert of grass, bare earth and salt flats"

The Bolivian Altiplano lies in the central part of the Andes, where the great mountain range of South America is at its widest. Here, the parallel lines of peaks – the western and eastern cordilleras – stretch far apart, leaving a vast intermontane basin 500 miles (800 km) long and up to 80 miles (130 km) wide. The land

within, enclosed by volcanoes and peaks, is a world of the unexpected. Instead of rugged slopes and valleys, it is filled almost to the brim with an immense level plain – a frigid desert of grass, bare earth and salt flats.

The Altiplano's smooth surface is the product of millions of years of erosion and deposition. After the basin was formed, huge quantities of eroded material from the surrounding mountains were carried into it by wind and water, accumulating as layer upon layer of sediment. As more and more layers were added, those strata beneath were gradually squeezed to solid rock. In time, the infilling reached high enough to swamp the basin's underlying topography. The top of the deposits currently stands at

The Salar de Uyuni is a huge salt flat on the Bolivian Altiplano. Early morning light falling near its rim softens the glare from the salt plain and picks out its only surface features – geometric ridges just a few inches high. Intense evaporation across the bed of a dried lake produced the salt deposits, which are more than 23 feet (7 m) thick in places.

Giant puya plants are welcome shelter for creatures in this treeless landscape.

12,000–12,500 feet (3,650–3,800 m), forming a strikingly flat surface, broken only by a few ridges, hills and volcanic cones.

Conditions on the Altiplano are inhospitable indeed. During most of the year, rain seldom falls, and in the thin mountain air, strong sunshine quickly evaporates away much of the moisture from the occasional summer storm in December and January. After sunset, the air cools fast, to below freezing, and sweeping gales can chill the land at any time. The windy, cold, desert climate is reflected in the thin, dry soil across most of the region and the sparse vegetation.

In the north around Lake Titicaca, the weather and the soils are slightly better, and people have long been able to make some sort of living from the land. Yet compared with the fecundity of the forests just to the east across the mountains, life has a difficult hold even here. Apart from clumps of bunch grasses, which retain old leaves around the edge as some insulation against the cold, there is little plant life save hardy cushion plants and cacti. On a few scattered hillsides live colonies of the giant puya – a distant, slow-growing relative of the pineapple, with enormous bayonet-sharp leaves. It blooms only once before dying, growing a flowering spike that towers 30 feet (9 m) above the ground.

The Desaguadero, one of the few permanent rivers of the Altiplano, flows south through this region. The sole outlet of Lake Titicaca, it is a sluggish waterway continually depleted by evaporation in the strong sunshine and dry winds. It crosses decreasingly fertile land, flanked on each far side by the walls of the Altiplano. To the east, the snow caps of the cordillera gleam in the sunshine, their rugged heights home to condors. The mountainous horizon to the west marks the Chilean border and is studded with volcanoes, including the mighty Sajama which rises to 21,463 feet (6,542 m).

Eventually, the Desaguadero reaches shallow Lake Poopo. Large flocks of flamingoes, egrets, herons, ducks and geese feed here, their abundance in sharp contrast with the empty lands beyond.

Most of the water that reaches the intensely arid section of the Altiplano south of Lake Poopo soon soaks into the dry soil or evaporates away. Only

The majestic Andean condor finds sustenance in the harsh world of the Altiplano. From its home in lonely and inaccessible crags on the surrounding mountains, this giant scavenger can glide, seemingly without effort, for great distances across the bleak terrain in search of carrion.

SALT LAKES

The water of Altiplano lakes can be too mineral-rich for many aquatic organisms. But salt-tolerant algae do thrive in some places, staining the water vivid green, brown or red, and nourishing other creatures. Here, the calm waters of Laguna Colorada are colored reddish-brown by algae. Water loss from evaporation, with little replenishment in the desert climate, has left a concentrated broth of mineral salts. The glistening bergs of crystalline salt were probably formed during past phases of rapid evaporation.

when rainfall has been heavy do streams flow for any distance. But evidence of wetter times is obvious in the landscape, literally blindingly obvious in a series of salt flats, or "salars." In brilliant sunshine, the glare from these perfectly level plains of crystalline salt is dangerous to unprotected eyes.

The largest by far is the immense Salar de Uyuni, a sheet of salt of a whiteness more pure than driven snow. Several yards in depth, it is more than 80 miles (130 km) from rim to rim. It owes its formation to a humid time when the Altiplano was almost covered in water.

Around 15,000 years ago, the meltwater from ice-age glaciers and snowfields in the mountains created two enormous lakes, one in the north, one to the south. After the ice age, the lakes gradually dried out, leaving today's remnants – Titicaca in the north and Poopo in the south. The Salar de Uyuni was also part of

"Strange lakes laden with salt exist in the volcanic landscape"

the southern lake's bed. Thousands of years of intense evaporation removed the overlying water but left its dissolved salts behind and continued to suck the ground beneath dry, bringing yet more salts to the surface. The same process continues today after occasional heavy rain storms spread a short-lived coat of water over the salar.

Strange lakes laden with salt but not yet dried out exist in the rugged, volcanic landscape that forms the southern fringe of the Altiplano. This scarcely inhabited arid zone of sand, gravel, banded rocks and bubbling geysers, lies at about 14,000 feet (4,250 m) above sea level. The chill is even more intense at this level, and few creatures are visible across the barren terrain. Only silent herds of vicuña, graceful wild relatives of the llama, wander the desert in search of scattered pockets of vegetation in this, South America's most singular and mysterious wilderness.

PEOPLE OF THE ALTIPLANO

The slightly more fertile north of the Altiplano is the homeland of the Aymara tribe, whose god Viracocha is said to have risen from Lake Titicaca to establish their ancient culture. Old traditions still remain in the lives of many modern Aymaras. These are reflected in the thatched adobe huts in which they live, in their herds of llamas and alpacas grazing the tough bunch grasses, and in the old, staple crops such as potato and oca tubers and quinoa grain that they harvest from terraced plots.

In a land of thin air, frosty nights, sun-baked days and fragile soils, life is unquestionably harsh, but the natives of the Altiplano seem to have evolved a physical hardiness to match their surroundings. Studies suggest that special features of their anatomy and physiology enhance the intake and passage of oxygen through the blood, an invaluable adaptation for toil at high altitude.

Llamas play a crucial role in the traditional life of the natives of the Altiplano. Hardy and dependable, llamas are beasts of burden as well as sources of wool, leather and meat. Even their dung is not wasted – it is dried in the intense sunshine and used as fuel for cooking and heating.

In spite of the harsh climate and thin soils of the Altiplano, tribal peoples have long made their home as farmers and herders in this uncompromising environment. The livestock they tend – llamas and alpacas – can find food even in volcanic deserts like this, where pasture is at best scant and coarse.

The potato probably had its origins as a crop plant on the Altiplano. Certainly it had been cultivated here for many centuries before it was discovered and introduced far afield by Europeans. Tolerant of dry soils and the bitter cold of night, it remains a staple crop of Altiplano people.

THE PANTANAL

"A maze of mingling habitats creates an extraordinarily diverse assemblage of life"

The lowlands known as the Pantanal form an enormous wilderness in the heart of South America. This flat, rain-fed region, laced with waterways and submerged beneath silvery sheets of floodwater for half the year, is one of the most evocative places in all the continent. The sweeping views and breathtaking sunsets are a backdrop for scenes matched only on the plains of Africa. Nowhere else in North or South America are so many animals

A mosaic of habitats including rivers, swamps, grassland and forests makes the Pantanal one of the richest wildlife havens in all South America.

of such variety so spectacularly visible as in the Pantanal.

One reason for the exuberance of animal life is the sparse human settlement of the region. Because of the damp land and annual flooding, large-scale cultivation of the land is impractical, and access remains difficult. Giant cattle ranches lay claim to the land, but they are stocked at low density, and ranching has done little to change the essential character of the landscape.

But it is the natural shaping of the terrain that makes the Pantanal special. It straddles the border between Brazil and Bolivia and covers more than 75,000 sq. miles (200,000 km²) of lowland tropics. It is, in effect, a vast basin almost encircled by higher ground. Some of these surrounding lands reach heights of over 3,000 feet (900 m), yet the Pantanal is only 500 feet (150 m) above sea level.

Long ago the depression would have been much deeper and perhaps filled with lakes, but for thousands of years rivers from the surrounding highlands have deposited sediment in the basin. Today that sediment is 260 feet (80 m) thick in places and has created a flat surface across which rivers flow in channels in the dry season. But in the wet season the rivers' excess water spills across the basin. All the rivers flow into the River Paraguay, which divides the Bolivian third of the Pantanal from the larger section in Brazil.

LAND OF
SEASONAL FLOODS

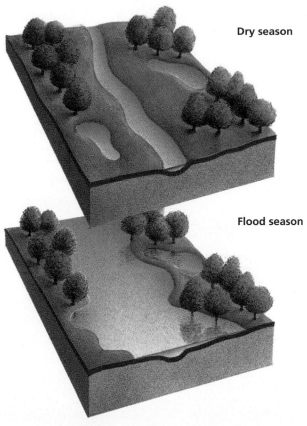

Dry season

Flood season

The landscape of the Pantanal changes with the seasonal floods. In the dry season, much of it is grassland broken by rivers and isolated pools. In the wet season, river levels rise and floodwaters spill across the land, leaving only islands of higher ground where trees and shrubs can survive.

Weed-choked pools in the Pantanal are favorite haunts of caimans. The largest are 8 feet (2.5 m) from snout to tail.

In geological terms the Pantanal plain is young, yet its newly formed landscape has been colonized by flora and fauna from all directions. Plants and animals from outlying regions such as the forests and swamps of the Amazon, the savannas of central Brazil and the thorn forests of the Chaco have all spread into the Pantanal, creating an extraordinarily diverse assemblage of life.

The Pantanal can support such variety because its landscape, though broadly level, is just uneven enough to create a maze of mingling habitats. The name Pantanal means "swamp." In truth, however, the region is a vast wetland mosaic of permanent rivers, pools, flood plains, patches of higher ground inundated only for short periods, and still higher ridges that are seldom, if ever, flooded.

The dry season in the Pantanal, when rains are few, runs from May to October. The floods gradually recede, and aquatic life becomes concentrated in rivers and around the diminishing pools. The permanent waters of the Pantanal are extremely rich in fish, ranging from small, colorful tetras to 30-inch (70-cm) heavyweights like the pacu. There are several species of piranha, most of which are plant-eaters, and large bottom-dwelling catfish, such as the surubim, which scour the mud for invertebrate prey.

Fish, especially piranha, are the mainstay of Pantanal caimans – the large crocodilians that spend much of their time around the swampy margins of the pools. Often seen wallowing and feeding alongside caimans are families of capybara. Resembling a giant guinea pig, this conspicuous rodent is the biggest in the world, with a maximum weight exceeding 130 pounds (60 kg).

The greatest spectacle of all is provided by the mingling flocks of water birds that crowd onto the dry season ponds to probe, sift and stab for prey. They include huge numbers

of egrets, herons, ibises and wood storks, as well as limpkins, which feed largely on snails, and lofty jabiru storks, symbol of the Pantanal.

On the flat ground surrounding the permanent waters, the soil becomes dry enough to support great fields of grass. These seasonal grasslands are the open savannas of the Pantanal, providing forage for fleet-footed pampas deer, marsh deer, peccaries (the New World equivalent of wild boars) and greater rheas (the "ostriches" of South America). Ants and termites make their nests on the grasslands, attracting giant anteaters and armadillos, and the numerous rodents and ground birds provide prey for maned wolves.

When danger threatens, many of the grassland creatures seek cover on the higher ground and ridges of the Pantanal. Generally free from annual flooding, these areas support thicker vegetation, with tall thickets in some places, groves of palms and even strips of dense forest, depending on the soil conditions. The trees and

palms form nesting and roosting sites for great numbers of Pantanal birds, among them egrets, cormorants, jabirus and hawks.

The rains come again in the wet season between November and April. The rivers draining into the Pantanal break their banks, and the grasslands soon become inundated, in places to a depth of more than 10 feet (3 m). Just

as the landscape of the Pantanal changes character during the flood season, so does its wildlife spectacle. Aquatic animals have the chance to spread. Now the dry land fauna must become more concentrated. Savanna and forest animals mingle on the scattered islands of high ground beyond the reach of the rising floodwaters in this watery wilderness.

A capybara with a large family to guard takes a wary look ahead for any dangers (right).

The biggest and most distinctive of the Pantanal's many water birds is the jabiru stork (below).

THE PATAGONIAN DESERT

"The last tapering segment of South America"

A cross the level plains that fill Patagonia, rain seldom falls. The soil is dry and the landscape flat and treeless, often without feature until it blurs into the distance. Winds sweep across the arid ground, fanning choking clouds of dust and obscuring the horizon. Nights are cold and winter is harsh here, for Patagonia, the last tapering segment of South America, stretches farther toward Antarctica than any other continental land mass.

Such a bleak land offers so little comfort for people, that across its vast 260,000 sq. miles (675,000 km²), the Patagonian Desert has no major cities, just a few towns and roads, widely scattered settlements and thinly stocked sheep ranches. The rest is desolation.

Patagonia consists, in essence, of mountain and plain. The spine of the southern Andes runs along its western edge, dividing the narrow, mountainous strip of Chile from the much broader Argentinan Patagonia to the east. With the towering peaks as backdrop, the land rapidly levels out to the east into a complex of plains and valleys. The terrain descends until it reaches a continuous line of cliffs fronting the Atlantic. In the north, the land drops in a series of broad steps. Farther south, the pattern is less regular, with evidence of volcanic eruptions in dark sheets of basalt and hills of crystalline rock.

The past also reveals itself in the series of broad, deep valleys – many with steeply walled sides – that cut east through the Patagonian plains. The rivers that flow through them now are too small to have made such a dramatic mark on the landscape. The valleys were probably carved by

Vast empty landscapes of browns and grays, their harshness accentuated by stark hills, characterize the Patagonian plains. Though increasing numbers of people have settled in Patagonia over the last 150 years, immense stretches remain lonely and untamed.

meltwaters pouring from the Andes during times when ice blanketed the heights much more extensively than it does today.

Giant glaciers ground down the flanks of the mountains then and gouged out basins at their feet. Today several of these basins are filled with large lakes. These lakes and the evergreen forests of southern beech that line the foot of the Andes are a haven for wildlife. Elegant waterfowl nest on the water's edge, and deer run in the woods in a scene of natural splendor very different from that of the desert close by. The rocky Atlantic coastline, with its rich waters

Ancient trees transformed into stone form these logs of petrified wood. Trees cannot grow on the arid plains today, but these dramatic fossils show that, long ago, the climate must have been much more humid.

and swirling flocks of seabirds, that rims the desert over to the east, provides a similar contrast with the arid somberness of the plains.

The Patagonian Desert exists because the Andes rob it of water. The westerly winds that prevail over Patagonia bring thick cloud to the slopes of the mountains. As the air mass rises toward the cold summits, most of the moisture condenses out and falls in a concentrated belt of heavy rain or snow. The last reserves are spent on the eastern slopes, watering the beech forests at their base, but leaving the air that crosses the plains dry and virtually cloudless.

Most of the Patagonian Desert receives less than 8 inches (200 mm) of rain per year. The combination of dry air and warm sunshine over the plains also means that evaporation is high, intensifying the drought. The rivers fed by rain and melting ice in the Andes steadily diminish in volume as

they cross eastern Patagonia through their oversized valleys. Many of them flow intermittently, leaving their watercourses empty for long periods with just a few saline pools to show that water runs from time to time.

Not surprisingly, the vegetation cover of the plains is poor, becoming worse with increasing aridity from north to south. In the region close to the Rio Negro (the river generally regarded as the northern limit of Patagonia), the landscape is one of open bushland. Thorny thickets up to 7 feet (2 m) tall stand widely spaced over bare soil. In sandy areas, grasses predominate. Farther south, where the plants have to survive with a minimal water supply, the bushes become ever lower and more scattered. In many places, the ground is covered only with rounded gravel.

Toward the Andes, the desert proper changes to a zone of semi-arid steppe, where bunch grasses take over. But whether scrub or grass predominates, the scene across plain and plateau is one of evenly spaced, knee-high uniformity.

In such a desert, it is all the more startling to find rebellious signs of animal life – the tracks of a snake, the dismembered remains of a scorpion, the prints of egrets around a drying pool. Like all deserts, the arid plains of Patagonia have their fauna, hardy creatures that have adapted to tolerate the dramatic swings between warm and cold, and can get by with low moisture intake.

One of the greatest problems for a desert animal, especially on the open plains of Patagonia, is the lack of cover in which to take refuge and hide. Successful adaptation to desert conditions reflects this problem. The guanaco and the lesser rhea, a large flightless bird, have evolved the escape strategy of speed. Both can run fast when danger threatens; indeed, the guanaco can outrun a horse. Similarly, the mara, a rodent

The towering pinnacles of the Paine Towers, part of the Patagonian Andes, are visible from far across the plains.

Prevailing wind

Heavy rain

Moisture evaporating from ocean

Desert in rain shadow of Andes

of the plains, has the long legs and body shape of a hare to help it race away from predators. The mara takes refuge at night in burrows dug in the earth. Several other animals depend more or less exclusively on the burrowing strategy to escape from danger, and have strong limbs and claws for digging. They include the pygmy armadillo, a type of rock rat and several species of tuco-tuco – compact, short-legged rodents that dare to make only brief foraging excursions above ground.

Predators of the desert, in turn, do their best to counter the defense strategies of their prey. The slender

DESERT IN A RAIN SHADOW

The Patagonian Desert lies in the rain shadow of the Andes. Air currents coming from the west are forced to shed almost all their moisture as they cross the mountains, leaving precious little for the plains to the east.

Patagonian weasel can pursue rodents into burrows, while foxes and the puma use a combination of stealth, ambush and speed to catch their quarry. The puma is fast and strong enough to hunt guanaco, as long as it can steal within range first, crouching as best it can behind low vegetation.

The most conspicuous hunters of the plains are the birds of prey, scourge of the tuco-tucos. Hawks and eagles employ the advantage of flight and acute eyesight to the fullest. Soaring high in the sunshine, they can scrutinize broad areas with little effort, the desert with its living secrets spread wide open beneath them.

THE TIBESTI MOUNTAINS

"Surrounded by blistering desert on all sides, this rugged terrain remains remote and wild"

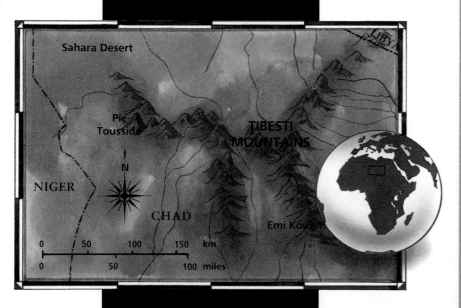

Sahara Desert

LIBYA

Pic Tousside

TIBESTI MOUNTAINS

NIGER

N

CHAD

Emi Koussi

0   50   100   150   km

0   50   100   miles

At the heart of the Sahara Desert lie the spectacular mountains of Tibesti. Visible from far away across the desert wastes, this isolated highland massif soars from rocky and sandy plains to heights of more than 10,000 feet (3,000 m) in a series of volcanic peaks topped with craters.

The Tibesti range covers an area greater than Switzerland, and its mountainous landscapes, though dominated by bare rock and desert dryness, are no less dramatic. From the summit of Emi Koussi, the tallest of the volcanoes, jagged crests and steep cliffs tinted with pink and violet stretch away to the north and west. Between them are broad platforms of land, some studded with curious rock towers and deep, dark ravines.

Surrounded by several hundred miles of blistering desert on all sides, this rugged terrain remains remote and wild. Once a hideout for raiders of camel trains, it has also long been a refuge and fortress for desert tribes fiercely resistant to outside control.

Though the Tibesti Massif lies within the hot desert, the mountains create their own modified climate. Temperatures are cooler at altitude compared with the desert plains, rarely rising to more than 68°F (20°C) by day, and dropping below freezing at night. Rainfall, though still sparse, is slightly higher than in the surrounding desert. Pockets of water collect in depressions, in the floors of craters and in the bottom of mountain valleys. Occasionally, heavy clouds build up over the peaks, sending stormwater through otherwise dry

A forbidding wilderness, the Tibesti Massif is a vast sandstone plateau, studded with the towering remains of giant volcanoes. The summits of the peaks are high enough for frost to form in hollows in the rocks at night.

> *"Tibesti was born of earth movements that uplifted ancient sandstone strata to form a plateau"*

valleys or "wadis." At these times, rock and mud debris is flushed into the valleys, and rainwater streaming from the massif can penetrate as far as 100 miles (160 km) into the surrounding plains of sand and stone.

The mountains, with their slightly more ample supply of water, are able to harbor more vegetation than the desert beyond. In addition to more than 300 species of small, drought-resistant herbs, grasses and bushes, the Tibesti has pockets of acacias and palms growing in its dampest sites. Surprisingly dense groves develop along some permanent streams.

Moisture and plant life support a simple community of animals. Seeds and other plant matter nourish desert rodents, including gerbils, spiny mice and gundis. Snakes and lizards are quite numerous, among them an insect-eating gecko that lives high in the mountains. Rock pools are visited for drinking and bathing by larks, sandgrouse and trumpeter finches. Doves, bulbuls and sunbirds live in the better vegetated areas, while white-crowned black wheatears snatch insects on the rocky slopes.

Among the most common larger predators are Rüppell's foxes, fan-tailed ravens, lanner falcons and tawny eagles. Barbary sheep, the hardy wild denizens of north Africa's mountains, are still fairly abundant in Tibesti, where they browse avidly on acacia leaves, flowers and even bark.

Nevertheless, Tibesti today is a predominantly dry, barren massif. Most of its surface is bare rock or rock fragments. But the area has not always been so arid. Global climatic cycles once brought more humid conditions to Tibesti. Higher rainfall in the past meant that Mediterranean oaks and pines grew on the mountains; the moisture nourished powerful rivers that carved much of the present terrain and dumped the eroded material across the surrounding lowlands.

Some of the rivers were big enough to reach the Nile and the Niger; others flowed into distant interior basins of the Sahara, where the water dispersed and evaporated. Numerous prehistoric rock paintings in the area show that antelopes and other animals more typical of the African savannas used to dwell around Tibesti, and the remains of crocodiles have been discovered in sediments at the foot of the mountains.

Evidence of a violent era much farther back in time exists in spectacular fashion in the relics of volcanoes high in the Tibesti Massif. Tibesti was born of earth movements that uplifted ancient sandstone strata

BIRDS OF THE DESERT

Few birds are so at home in deserts as the family of plump, dovelike fowl known as sandgrouse. All the 16 species live in arid and semi-arid environments in Africa, Arabia and Asia. Two of these haunt Tibesti – the crowned sandgrouse and Lichtenstein's sandgrouse. Capable of fast, direct flight, they can travel large distances across the desert to visit waterholes and areas where seeds – their principal food – are plentiful.

The male sandgrouse has a remarkable adaptation to desert life. He has specialized belly feathers that can soak up water from pools rather like a sponge. This means he can bring water from distant waterholes back to the nest so that thirsty chicks can sip the precious moisture.

Grains of sand driven by strong winds abrade and undercut exposed rocks in parts of Tibesti, creating some spectacular and bizarre landforms. Water from occasional rainfall also plays its part in the shaping of the landscape, though less so now than in humid eras of the past.

The lanner falcon finds secure, sheltered nest sites on the rock faces of Tibesti. This powerful predator is the scourge of smaller birds, which it generally snatches on the wing. It also preys on small mammals, such as gerbils and spiny mice, and catches large insects including locusts.

to form a plateau. Probably at the same time, molten rock, or magma, from below the Earth's crust pushed its way upward through weaknesses in the plateau. Over millions of years, these volcanic outpourings built sheets and cones of basalt, tuff and lava on top of the sandstone. Today, most of the higher parts of Tibesti are built of volcanic rock.

At 11,204 feet (3,415 m), Emi Koussi is the highest peak – and one of the most dramatic. Its sides slope up evenly all around, then drop away suddenly into a vast, cliff-ringed crater roughly 12 miles (20 km) across and 3,000 feet (900 m) deep. The crater evokes a lunar landscape, but it was created not by a falling meteorite, but by eruptions from within.

About five million years ago, an enormous volcano began to develop.

Lava from a great subterranean chamber of magma poured from its vent and cooled on its flanks, building up a massive cone far higher than it is at present. Eventually, however, the emptying magma chamber could not support the weight of the cone above. The top of the cone began to collapse, probably in gradual stages, leaving the unmistakable imprint of volcanic self-destruction – a giant circular depression known as a "caldera."

Trou au Natron is a similarly spectacular caldera, just as deep and about 5 miles (8 km) in diameter. But its genesis appears to have been rather different. The volcano from which it originated did not build a tall cone, and its eruptions appear to have been more explosive. Indeed, it is believed that the caldera was

blasted into existence by a series of titanic explosions that blew rock away and caused the remainder to collapse. Chunks of debris weighing many tons have been found 6 miles (10 km) from the caldera.

Volcanic activity on such a scale has long ceased, but Tibesti is by no means completely quiet today. Gases and hot water still stream and bubble out of vents in Trou au Natron, evidence that magma remains not too far below. The mineral-rich springs have left blinding white deposits of carbonate salts across the crater floor. Elsewhere in the Tibesti heights, there are several other places where hot springs emerge, along with fumaroles, geysers and sputtering mud pools. All contribute to the savage beauty of this extraordinary mountainous island in the midst of the desert.

THE BALE MOUNTAINS

"The single largest area of high ground in all of Africa"

Ethiopia is a mountainous land. In a continent that is dominated by hills, valleys and plains, Ethiopia's rugged core stands in striking contrast. One of its loftiest and least explored sections lies toward the southeast: the frosty heights of the Bale Mountains.

These mountains form part of an upland mass cut off from the rest of

Red-hot poker plants make the most of the strong sunshine at high altitude on a grassy stretch of the Bale Mountains.

the Ethiopian Highlands by the northern trench of east Africa's Great Rift Valley. The mass is further cut in two by a river valley, the upper course of the Wabe Shebele. The Bale massif lies south of this divide, forming the single largest area of high ground in all of Africa, with at least 400 sq. miles (1,000 km²) topping an altitude of 11,000 feet (3,300 m).

From the open mountaintop, the land falls away on all sides in steep escarpments and narrow valleys, cloaked in greenery. The vegetation is fed by heavy rains deposited on high by winds direct from the Indian Ocean – from March to October, it rains or hails on the summits almost every day. The great Harenna Forest spanning the southern flank is one of

the biggest stretches of mountain forest left in Africa. It continues downslope until it blends with acacia woodland and savanna on the lower, drier contours.

Tectonic wrenching of the Earth's crust and upwelling magma from far below combined to push Bale and the rest of the Ethiopian Highlands up out of the African plains long ago.

The huge bulge of rock so created was deeply cracked and fissured by the stresses. Other forces widened these fault lines, breaking up the dome and encouraging deep valleys to form between the mountain blocks. Immense splitting on a continental scale then rent the Great Rift Valley into existence, severing the highland dome completely.

As world climate repeatedly cooled during ice ages, ice caps formed on the Bale Mountains and brought tundra conditions to the slopes. Today, though the ice-age chill has long receded, nightly frosts on the tops split shards from exposed rock and create upheavals in the soil. Forests can grow on the slopes, but the heights remain too cold.

Partly because of its broad expanse of cold land, offering few sheltered valleys, the Bale massif has remained relatively free from human influence. While the rest of the Ethiopian Highlands has been severely degraded by the pressures of farming and settlement, the natural landscape of the Bale Mountains remains more or less intact. In an effort to keep it so, a large section of the massif has recently been set aside as a national park.

From south to north, the park traverses the full range of habitats on the mountain slopes. Thick tropical forest of yellowwood, stinkwood, African olive and other trees on the lower slopes blends through clouds and mist to a belt dominated by reddish-flowered hagenia trees. Higher still, the landscape takes on a strangeness unique to high east African mountains. St. John's wort and heather predominate, but they are not the modest plants familiar in northern Europe. The varieties here grow to the size of trees and are draped in moss and lichens in the foggy, damp conditions.

A pattern of shrubby trees and glades gradually turns to open terrain across the highest ground. In places, especially toward the north of the park, fairly dense grass grows, dotted with colorful red-hot poker plants. But the high core of the massif, the Sanetti Plateau, is a bleak, windswept

"St. John's wort and heather grow to the size of trees in the foggy, damp conditions"

Simien jackals live in close-knit packs on the Sanetti Plateau. The pack works together to provide pups with food and protect them from danger, such as marauding birds of prey. Here a young jackal howls an alarm call to alert fellow pack members (right).

Giant lobelias thrive on the bleak Sanetti Plateau – the coldest, highest section of the Bale Mountains (left). At night, each cluster of spiked leaves closes up to protect the developing shoot of emerging flower heads from freezing conditions.

zone, reminiscent of moorland. More than 11,500 feet (3,500 m) high, the plateau is a gray-brown mixture of tussock grass and aromatic herbs such as lady's mantle and sage.

Most plants are ground-hugging to cope with the cold and the wind, but the mountains have another curiously oversized plant. Here and there, giant lobelias erupt from the diminutive ground vegetation of the plateau, with their thick stalks, spiked clusters of leaves and towering flower stems reaching up 16 feet (5 m).

On the flatter portions of the plateau, bogs develop, with small pools and lakes that are attractive to wildfowl. Some of them are birds familiar in other parts of Africa, like Egyptian geese, yellow-billed ducks and wattled cranes. Others, including white and black storks, drop into the high-altitude lakes for a rest during migration. But there are also birds that occur only in the heights of Ethiopia. The blue-winged goose spends all year on the pools and moors, its loose, thick plumage providing insulation from the cold.

The plateau and the grasslands are home to other animals unique to Ethiopia. Stately mountain nyalas, a type of antelope, have their stronghold in the north of the park, where herds a hundred strong sometimes congregate. They feed on the grasses and herbs by day, but retreat among the St. John's wort groves at night, where temperatures are milder. Simien jackals stalk the Sanetti Plateau, where they feed largely on rodents.

Bale also has its own species of hare, Stark's hare, and harbors a unique bird, the Rouget's rail. Indeed in the Ethiopian Highlands, and Bale in particular, a high proportion of animal species are endemic – they occur only here – a measure of the extreme isolation and unusual nature of these rugged mountains.

THE SERENGETI

> *"Immense flat plains of grass where large grazing animals exist in unparalleled numbers"*

Serengeti National Park

Serengeti-Mara ecosystem

There is nowhere on Earth quite like the Serengeti. On this, the most famous of the east African plains, the herds of wild mammals easily outnumber those of any other place. The grazing animals can amass in such concentrations that they hide the ocean of grass beneath them. Yet, at other times, the same expanse of grass

stretches seemingly empty as far as the horizon, abandoned by the herds whose instincts command them to wander with the seasons.

The Serengeti is a complex of savanna habitats, spread over a broad surface of volcanic soil, 5,000 feet (1,500 m) above sea level. The Serengeti National Park protects 5,700 sq. miles (14,750 km²) at its heart, but the full extent of the Serengeti takes in nearly twice this area. From Lake Eyasi in the south, flanked by highlands to the east and stretching toward Lake Victoria in the west, it reaches north to incorporate the Masai Mara Game Reserve across the Kenyan border. Immense flat plains of short grass in the southeast blend into areas of longer grass and

The Serengeti is one of the greatest *remaining treasures of wilderness in Africa. Here, large grazing animals, such as the elephant for which the African plains are so well known, exist in unparalleled numbers. Unhindered by fences and farms, they can wander across the vast open grasslands and woodland savanna.*

Rest takes up *much of a lion's day, especially at times when prey is abundant and life is easy. Females do most of the hunting. Crouched low in the grass, they stealthily approach prey before springing to attack.*

Rocky outcrop

Short-grass plain

Acacia tree

Ash soil

Hardpan layer of calcium carbonate

A LAND OF GRASS

Volcanic soils full of compacted ash underlie the vast expanses of grassland in the southeastern Serengeti. Grasses, with their fine, shallow roots, take hold readily in this soil, but bigger tree roots find it difficult to penetrate. They are further impeded by a rock-hard layer of carbonate deposits.

The acacia is one of the few trees whose roots do manage to break through the hard layers and establish themselves. Such trees stand out like sentinels on the grassy plain (left).

then to the north and west turn into a broad swath of woodland savanna, where thorny acacias compete for space with the surrounding grass stems. Sometimes grasses take over, leaving acacias scattered in their midst; elsewhere the thorn trees crowd into tall thickets, with the shorter plants clumped patchily underneath.

Though the acacia woodlands bring variety to the landscape, greatly increasing its wildlife diversity, grassland remains the key to the Serengeti. Grass, and the small herbs

that grow among it, is the mainstay of the vast herds. It is food for some 1.5 million wildebeest and another million zebras, gazelles and buffaloes, and myriad other creatures, from termites to ostriches. The abundance of grass in the Serengeti, particularly in the almost pure-grass plains of the southeast, seems to be a product of the soil rather than the climate.

The plains themselves were built largely of wind-blown ash carried by prevailing winds from volcanic highlands to the east. Some of that ash formed hard concretions, with the grains bound so tightly together that tree roots now have difficulty finding a route through the soil.

The surface of the ash soil is highly porous to water. Rain passing quickly through the surface layer has tended to leach the calcium-rich deposits, dissolving out the mineral content and then depositing it again as a hard barrier of calcium carbonate 3 feet (1 m) or so below ground. Trees also have difficulty establishing themselves in these conditions because their roots are impeded, so space is left free for the short, flexible roots of grasses. The parts of the Serengeti closest to the parent volcanoes are those where the barriers to tree growth are at their most accentuated, and

grasses face the least competition for space.

Grass is wonderfully resilient, tolerant of fire damage, trampling and grazing. So long as grazing pressure is not too heavy, it can grow afresh time and again after its leaves and stems have been clipped away. This is because the plant's growing points are at the base and its oldest parts at the top, pushed ever upward by new growth below.

But even grass has its limits. Overtaxed pasture soon loses its productivity, especially if dry seasons rob the soil of precious moisture. Changing weather conditions in the Serengeti force the largest of the grazing herds to undertake epic migrations in search of fresh pasture. Their dramatic movements affect many of the other animals that share the Serengeti with them.

Wildebeest are the greatest of the mass migrants. In April, legions of them fill the southern short-grass plains. Many of the females have given birth only two or three months before and are busy nursing their calves. Yet within a month all will feel the stirrings of ancient wanderlust. The rains start to dwindle in the south, and the grassland begins to lose its verdant tint. Soon it will be dry and exhausted. Gradually, the animals become restless and start to move. Small herds walking in line join with others heading northwest, and as paths converge, the bigger herds coalesce.

Before long the wildebeest have amassed to form giant dark columns, streaming with singular purpose across the open plain into the longer grass zones and beyond. From the vantage point of a smooth rock outcrop – these large boulders rising like islands from the grass are a feature of the plains – half a million animals may be in view stretching for several miles to the horizon. As the wildebeest stream around the outcrop, noise fills the air and veils of dust settle around the small animals – perhaps agama lizards and hyraxes – that dwell on the rock.

Among the wildebeest and their calves, others join the procession toward better pasture. The stripes of zebras and the tawny coats of gazelles are readily visible. Though they are all grazers, there is little competition for food between the different animals. When the migrating herds pause to rest and feed, different feeding preferences are revealed. The zebras can cope with the fairly coarse tops of long grass; their chewing exposes more tender parts below. Wildebeest clip the sward further, leaving the lowest, newest growth for Thompson's and Grant's gazelles.

Plains zebras *stoop to graze on long grass in the Serengeti (below). Their stomachs are adapted for digesting the tops of long grass – the oldest, toughest parts of the plant.*

Wildebeest face many dangers on their annual migration. Perils such as predators, drowning and injury during stampedes claim an estimated 40,000 victims each year.

The plains through which the migrants pass are the permanent homes of other animals. Some are also grazers, but do not take part in the great migrations. Buffalo, hartebeest and topi tend to stay put and nibble what they can during the dry season. For some other animals, finding sustenance is hard work for most of the year, but the weeks when the herds are in the locality are a time of bonanza. Packs of hyenas living in the center of the Serengeti may be drawn 30 miles (50 km) or more from their denning area to take advantage of the glut of fresh meat.

The young herbivores, struggling to keep up with the herd and easily separated when the adults are frightened into stampeding, are among the most vulnerable prey. But adult animals are also picked off: sprightly gazelles often fall prey to cheetahs, wildebeest to hunting dogs and zebras to lions. If the supplies of food for such predators were constantly abundant, there might be many more of them. But lean times during the rest of the year, when they

"Dark columns of half a million migrating wildebeest stream with singular purpose across the open plain"

are left to hunt scattered prey – often consisting of smaller mammals and birds – serve to limit their numbers.

The wildebeest continue marching until they are well into the woodland savanna, some moving west almost to the flood plain of Lake Victoria. Here they may linger for several weeks in the company of impalas, elands and giraffes before continuing into the northern Serengeti where the dry season is relatively mild and good

pasture remains abundant. Their journeys force them to cross rivers, where the healthy size of the resident crocodiles shows the benefit gained from several weeks of easy hunting.

For many thousands of wildebeest, the long journey north does not finish till they reach the Masai Mara reserve. To do so, they have to swim across the Mara River. There are only a few suitable crossing points on the river, and great herds mass at these bottlenecks, nervously waiting for the first bold individuals to plunge into the water. The rest follow in a chaotic scramble. Having swum the breadth of the river, they clamber frantically up the opposite bank which is usually steep and muddy. On every crossing, the water becomes strewn with animals that have drowned or been trampled in the panic.

Wildebeest linger on the rich Masai Mara grasslands until October when the wet season returns. Then the wanderlust revives. As migratory birds escaping from the approaching winter in Europe and northern Asia arrive in the Serengeti, so the vast herds start to trek south again. By December those that have survived the hazards of the march are back in the southern plains, feeding on the fresh grass revived by the rains.

Conservation of the natural riches of the Serengeti has not been without problems. Some damage to the park and reserves has resulted from settlement nearby, and poaching has put severe pressure on some wildlife, such as the critically endangered black rhinoceros. But, at least for the time being, the Serengeti has few problems of competition for pasture between wild animals and livestock. No fences have been erected across the region to halt the mass seasonal movement of the animals. Once much more widespread in Africa, such wildlife spectacles have all but faded into distant memory. In the Serengeti, they live on.

A Masai warrior displays his weapons outside a typical dwelling (right). All men become warriors after initiation into manhood. Once feared for their combat skills and cattle-raiding forays, the warriors' chief role now is to manage herds and ward off predators.

A Masai woman decorates an ox hide. Cattle are central to the subsistence, material life and customs of the Masai. Their blood, milk and meat provide food; their skins, horns and bones are used in many objects – decorative and practical; and their hides are also used to protect the mud huts in heavy rain.

The livestock herds of the Masai share the grassland plains with wildlife (right). Though overgrazing does occur more often today, the traditional way of life of the Masai is well suited to the fragile savanna. When a pasture becomes thin, each village takes its cattle and belongings to a new site.

THE MASAI

Grassy plains in Kenya and Tanzania are the homeland of the Masai. These people are renowned throughout Africa for their dignity and valor, for their adherence to tradition in lifestyle and values, and for their ornate crafts.

The Masai live by raising cattle, goats and sheep on the immense pastures that surround their settlements. Life revolves around the livestock and their need for food, water and protection from wild predators. The Masai move their villages with the seasons to make sure the herds have enough fresh grass.

The simple dwellings of the Masai are neatly in tune with the environment. Built from a framework of sticks packed with leaves and plastered with mud and dung, they are quick to make and proof against strong winds in the open terrain. The dung in the plaster is also an effective deterrent against termite attack.

TAI NATIONAL PARK

"The remote forest interior offers a lifeline to the threatened wildlife of west Africa"

West Africa's largest remaining stretch of virgin rainforest is contained in the Tai National Park. Beneath its sunlit canopy, in the eternal shade of the open floor, the forest feels hauntingly tranquil and all-enveloping. In every direction, trunks rise from the gloom far into the mass of greenery above. Vines and creepers hang from them

A curtain of greenery encloses the rivers and pools of the Tai rainforest. It droops right down to the water where creatures such as pygmy hippos wallow (right).

like stiff, twisted ropes. Quiet and yet burgeoning with life, the forest with its untold secrets can feel as stifling as the still, humid air.

Tai remains a refuge for animals and plants hunted and harvested out of existence in much of west Africa. Valuable hardwood trees stand in Tai's remote interior, and elephants, though hard-pressed by poachers, continue to roam across its floor.

Extending over some 1,350 sq. miles (3,500 km²), Tai National Park lies in the southwest of Ivory Coast, between the Sassandra River and the Liberian border. Its forest covers a lowland area of gently sloping plains and scattered hills. Mount Niénokoué, the highest point, is an isolated granite dome rising to 2,044 feet (623 m).

The luxuriance of Tai's vegetation is encouraged by conditions common to all true rainforests – constant heat and humidity. Temperatures average 80°F (26°C), with only minor variations through the year. Rainfall, though not as high as in central Africa, is still copious when compared with the rest of the continent. The period from December to February tends to be drier than the rest of the year, but no month is without rain.

Uninhibited by seasonal change, the plant life of the forest is in a constant state of growth. Flowers and fruits are ever present. The great bulk of tree foliage lies in the canopy, where the crowns of medium-sized species mingle. Scattered bigger trees, with massive columnar trunks often buttressed at the base, push unbranching through the canopy to emerge like sentinels in the sunlight above. The diversity of plant life is wondrous – there may be 600 different species of trees alone.

It is in the canopy and emergent trees that animal diversity is also at its most staggering. Here teem innumerable beetles, bees and spiders, lizards, snakes and tree frogs. The calls of birds flocking around a fruiting tree echo from on high, breaking the tranquillity of the forest.

The Diana monkey lives high in the forest trees. Here, in the cooler hours of dusk and dawn, it forages for buds, leaves, insects and even bird's eggs, as well as fruit. Very active and noisy troops of up to 30 of these monkeys sometimes race through the trees.

"The diversity of life is wondrous – there may be 600 different species of trees"

FOREST SOIL SYSTEMS

Rainforest soil is generally poorer than that of a temperate forest. Heavy rainfall leaches soil nutrients beyond the reach of roots. But fallen leaves decompose rapidly, and trees spread their roots just under the litter to reabsorb as many nutrients as possible. Buttresses give these shallow-rooted trees extra support.

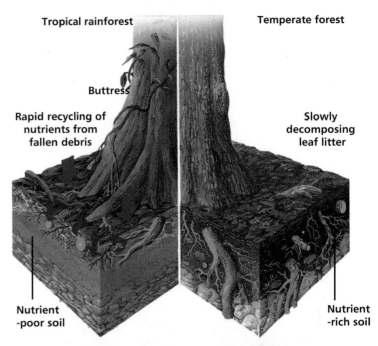

Tropical rainforest

Temperate forest

Buttress

Rapid recycling of nutrients from fallen debris

Slowly decomposing leaf litter

Nutrient -poor soil

Nutrient -rich soil

Ripe fruit can attract visitors from some distance across the forest – it is not unusual to find parrots, fruit pigeons, hornbills and barbets feeding alongside one another. Several different species of monkeys may also converge, chattering and feeding, at the same site, among them red, green and pied colobus monkeys, sooty mangabeys, and vividly patterned Diana monkeys.

The ground layer of the Tai forest, with its thin mantle of shade-adapted vegetation, its fallen trees surmounted by moss, ferns and fungi, and its dark recesses at the bases of tree trunks, seems a distant world from that of the canopy suspended 100 feet (30 m) above. Most canopy animals fear to venture from their lofty home, while many of the creatures found at ground level cannot climb high. At night, many of the latter emerge from hiding to browse on low-growing plants and sort through the leaf litter on the forest floor. Among the rare animals for which the park provides refuge are the giant forest hog – the biggest of all wild pigs – and the pygmy

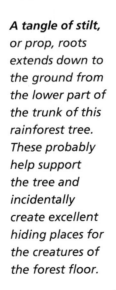

A tangle of stilt, or prop, roots extends down to the ground from the lower part of the trunk of this rainforest tree. These probably help support the tree and incidentally create excellent hiding places for the creatures of the forest floor.

hippopotamus – the comparatively diminutive cousin of the hippo.

A few predators such as leopards and golden cats hunt for prey both on the ground and in the trees. These cats are good climbers and can make surprise ambushes by leaping down onto prey. The rare but conspicuous chimpanzees also roam from forest floor to canopy. Troops of chimps tend to feed by day, sifting through

ground debris for edible items, chewing leaves in the understory, or climbing high to pluck fruit – their favorite food – from the canopy.

For the chimpanzee and other threatened wildlife of west Africa, the remote interior of the Tai forest offers a lifeline, a precious chance of survival. So long, that is, as the logging and poaching that threaten its fringes can be kept at bay.

THE OKAVANGO DELTA

"The biggest inland delta in the world and a glittering jewel in the heart of southern Africa"

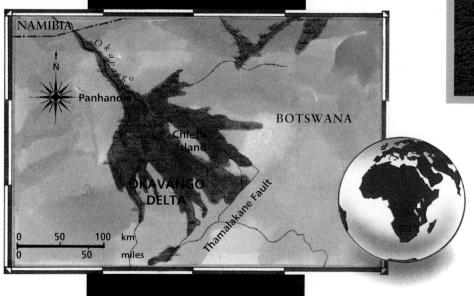

The Okavango River is doomed to perish. As it heads south through an ever-drier landscape of sand, thornbush and pallid grasses, little replenishment arrives to counter the toll of evaporation. Unlike most desert streams, however, the Okavango refuses to die a quiet, shriveling death. It ends, instead, in a blaze of glory. While it still has vigor, the river spills its waters into a giant, spreading fan. The radiating

waterways feed a maze of shallow backwaters, lagoons and swamps, bringing sparkling vitality to the desert plain. The river's death-throes have created the biggest inland delta in the world and a glittering jewel in the heart of southern Africa.

From the west, in the ocherous expanse of the Kalahari Desert, the delta lines the horizon with distant, beckoning turquoise. The blue of the water blends with the green hue of thick vegetation – crowded stems of reeds, sedges and grasses and luxuriant swaths of trees and palms. With good reason, the Okavango is likened to an oasis. Not only does it support a wealth of permanent wildlife, but it also draws thirsty and hungry wanderers from far and wide, especially when receding floodwaters leave green pastures in their wake.

The Okavango Delta lies on an immense sandy plain within an ancient inland basin. The waters that feed the Delta flow into the basin from the highlands of central Angola and head steadily southwest for 700 miles (1,100 km) into northern Botswana. There, earth movements have created a system of geological

Red lechwe feast on new shoots springing from the flood plain grasslands of the Okavango Delta. A fertile, well-watered haven in the midst of the desert, the Delta is one of Africa's richest wildlife areas.

faults that seem to have remodeled and interrupted the water's path. First, the Okavango River is channelled between two parallel faults, where it forms the so-called Panhandle of the Delta. As it emerges from these, it meets further fault lines that help to break up its flow into divergent channels. A few of these continue through the Delta and beyond until they meet the Thamalakane Fault, which has created a sand-covered rock barrier some 125 miles (200 km) long. The remnant streams change direction and combine as they follow the foot of the natural dam, before passing through a breach in the fault as a single, ever-dwindling waterway.

As the Okavango River spills out of the Panhandle to form myriad smaller, shallower streams, so it drops most of its sediment load. An estimated two million tons of sand and silt is dumped into the Delta every year. Much more could have been deposited at times in the past when the river was larger and probably ponded back by the Thamalakane barrier. In time, as the stream channels have repeatedly

> *"The yearly flood tide has a dramatic impact on the landscape and ecology of the Delta"*

become blocked by sediment and been forced to find new courses, this has caused the Delta to build outward as well as upward, ramifying into the complex fan-shaped pattern of wetland that covers more than 6,000 sq. miles (15,500 km²) today.

This wetland is an intricate mosaic of reedbeds, swamps and areas of open water, dotted with patches of higher ground where trees and other dry land vegetation can grow. Moreover, it is a changing mosaic.

Oxbow lakes – old meander loops cut off from the stream flow – testify to the transitory paths of water passage as erosion and deposition continually rework the surface. In recent times, one entire channel has largely dried up because its input has been diverted elsewhere in the Delta.

Islands and their vegetation change character as shifts in water levels expose or submerge more than before. Reeds trap sediment, so building up ground beneath them, and even Delta animals can bring about change. The mound-building habit of termites helps to establish islets, while the movements of hippopotamuses and elephants can break up banks and wear new water channels through reedbeds.

Toward the south of the Delta, the islands are larger, the land drier and the permanent swampland less extensive. This marginal zone of the Okavango is even more changeable. It has broad areas that are clothed with grass for much of the year but inundated when the Okavango waters make their small but influential annual rise. At this time the total wetland area can increase by some 2,300 sq. miles (6,000 km²).

The yearly flood tide in Okavango may be a sluggish event – the advancing high waters take about five months to pass from the Panhandle to the southern flood plain – but its impact on the landscape and ecology of the Delta is dramatic. As the flood-tide passes slowly through the different zones of the Delta, it replenishes oxygen in stale waters and redistributes nutrients. It also defines cycles of vegetation growth and provides the triggers and checks that govern the lives of the Delta's wild inhabitants.

The Okavango Delta has its rainy season from November to March, but the angry skies and scattered showers add little to the water levels. Real change begins to come toward the end

Aquatic plants thrive in the tranquil backwaters of the Delta. Only a few of the 80 or so species of fish in the Okavango chew at the leaves of water plants, but the red-breasted bream is one that can make short work of young lily leaves such as these.

AN ANNUAL FLOOD TIDE

Every year the Okavango River, fed by heavy rainfall in its northern reaches, brings an enriching flood tide to the Delta. Rising waters spread out across flat grasslands to the south, greatly increasing the Delta's size. They also stir up and spread out nutrients from mud and soil and re-oxygenate stagnant pools.

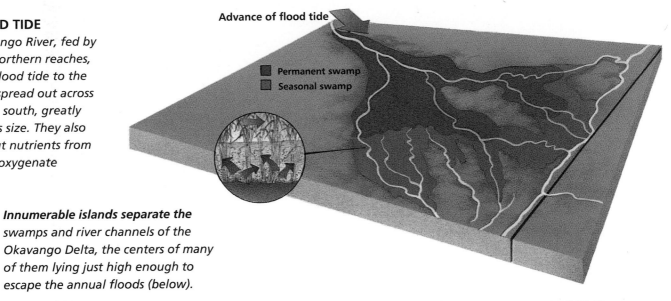

Advance of flood tide

Permanent swamp

Seasonal swamp

Innumerable islands separate the *swamps and river channels of the Okavango Delta, the centers of many of them lying just high enough to escape the annual floods (below).*

The saddlebill stork is a frequent visitor to the southern Delta during the flood season. As it wades through the shallows, it keeps its eyes firmly fixed on the water, ready to lunge with its bill at any sign of movement.

of the season, when more sustained rainfall across central Angola feeds the river downstream. The swollen river races first into the Panhandle, where it cuts its path through more or less continuous swamp dominated by tall papyrus stems and willowy palms.

As the river rises and washes over its banks, the swamps become replete with well-oxygenated water, in which aquatic plants and invertebrate animals thrive. Newly flooded land at the swamp margins becomes calving ground for sitatungas, the specialist antelope of the swamps, and huge numbers of migratory catfish wriggle into the shallows to spawn.

Large animals have difficulty moving through the swamp – the dense papyrus impedes them and the loose, spongy rootmats give little support underfoot. The sitatunga has large hooves that splay out as it treads over the mats, spreading its weight. Few predators can pursue it into its watery haven, although it does fall victim to the big crocodiles that lurk by the river.

As the flood tide crosses out of the Panhandle, it spreads over the much wider area of the Upper Delta. Yet there is still enough water in the swollen channels to flush through the swamps here, rehydrate drying reedbeds and lap up the sides of islands. Papyrus no longer dominates

the vegetation. Instead there is a mixture of reeds and sedges in water of varying depth, with high grass on flood plains and belts of lofty trees fronting the islands. Lazy waterways branch from the main channels, meandering sometimes into tranquil lagoons, where hippopotamuses spend their days and fish eagles, reed cormorants and otters hunt for fish.

Across such a habitat, flight is a tremendous asset, and it is little surprise that airborne creatures are among the most prominent of the Delta. Bats scour the waterways and dart over the reeds at dusk, snatching night-flying insects such as moths and mosquitoes. By day, the swamps are alive with bees and dragonflies, kingfishers and herons.

Thickets of water fig covering islands in some of the larger lagoons provide food for doves and parrots and harbor boisterous mixed colonies of storks, ibises, herons and egrets. Breeding in the colonies is timed so that the chicks hatch after the high waters have subsided. Fish and other prey are easier to catch when they are concentrated in shallower water.

For the fish, on the other hand, the flushing flood stirs up nutrients and opens up new foraging sites as the level rises. Many of the colorful cichlid fish of the Delta spawn during

high water so that their young can take advantage of the good feeding.

The flood tide still has not reached the southern margins of the Delta, where the flood plains remain covered with grass. It is dry season in the nearby Kalahari, and these precious Delta pastures have attracted grazing animals not just from the fringing acacia woodland but also from a large surrounding area of northern Botswana. The Delta now has peak numbers of zebras, buffaloes

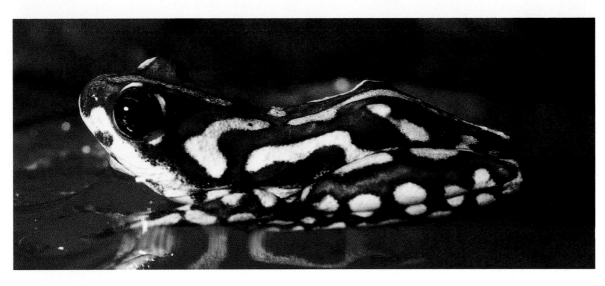

The tinkling calls of painted reed frogs add to the chorus of nocturnal life in the swamps. Male frogs are at their most voluble in the mating season from September to April when vying for the attention of females.

and elephants. They are accompanied by herds of wildebeest, giraffes, sable antelopes and impalas. Warthogs and chacma baboons are common, and the glut of game provides easy pickings for lions, hyenas and hunting dogs.

But the advancing tide front is creeping through the last stage of its journey. Soon it reaches the grassy plains and slowly spreads a sheet of shallow water across the flat terrain. Spawning fish wriggle among the drowned blades and, as the flood water becomes enriched with nutrients – from dead insects, dung, seeds and rotting grasses – many more will arrive, pursued by hunters such as the African pike. Dormant lily bulbs sprout from the submerged soil as they and other aquatic plants take their turn from the grasses. The new shoots provide food for geese and for the red lechwe – another of the Delta's characteristic antelopes.

Though most of the mammals are pushed from the flooded grasslands of the southern Delta, many retreat only into the bands of woodland, from where they can still emerge to slake their thirst from the fresh supply of water. The big nomadic herds of grazing animals tend to linger around the edges of the flood, concentrating on the grassy remnants before heading out into the wider Kalahari. In a few weeks, the floodwaters start to drop once again, and the tidal cycle of the great Okavango Delta turns inexorably on.

Hippopotamuses loll in lagoons in the deeper waters of the Delta by day. At night they emerge onto islands to crop the grasses. Their well-worn paths from water through swamp to dry land provide easy passage through otherwise impenetrable reedbeds for various creatures, including crocodiles.

PEOPLES OF THE OKAVANGO DELTA

Tribal villages lie scattered throughout the Okavango Delta, beside river banks or on sandy islands. People have lived in the area for thousands of years, drawing on the abundant food resources of this extensive swampy wilderness.

For all, life revolves around the waters. The rivers are places to drink, bathe and fish. They are the principal travel routes, and their annual floods renew the fertility of the soil for cultivation and for grazing animals.

The different tribal groups present in the Okavango – the River Bushmen or Banoka, the Bayei, and the Hambukushu –

share many of the same techniques for subsistence in the Delta. In addition to farming, people gather edible plants and hunt for meat in the swamps and woodland. Long ago, River Bushmen perfected the use of concealed pits dug along trails to trap game. They fish, using nets, hook and line, traps and spears. In shallow water, Okavango fishermen sometimes poison fish by throwing in dried and ground toxic bark from a local tree.

Travel is generally by shallow dugout canoe, carved from a single tree trunk and propelled either by pole in the shallows or by paddle in deeper water.

A small tribal village in the heart of the Okavango shows the neat construction of Delta dwellings, built from swamp vegetation. The grass-thatched huts are surrounded by fences of long, sturdy reeds carefully bound and trimmed. Most villages have cultivated plots and cattle pastures nearby.

Bayei women use elegant, funnel-shaped baskets for catching fish (below). After placing the baskets side by side in the water, the women wade toward them from upstream, flushing fish into the traps.

All the peoples of the Okavango are skilled at obtaining the Delta's most abundant protein source, fish. Here, a River Bushman displays his catch, all snatched from the water by spear (left).

THE SKELETON COAST

"A desert land of parched rock and shifting sands fronting a fogbound sea"

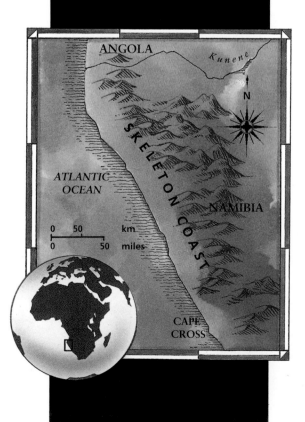

Few places in Africa evoke such an image of desolation as Namibia's infamous Skeleton Coast. The coast remains, to this day, lonely and forbidding – a desert land of parched rock, gravel and shifting sands fronting a treacherous fogbound sea. Its skeletons are the broken bones of foundered ships as well as the remains of long-dead castaways found half-buried in the sand. Yet there is life in this barren landscape, life that is surprising and ingenious in its adaptations to the environment.

The Skeleton Coast lies in the north of Namibia, running from Cape Cross up to the frontier with Angola. Most of it lies within the Skeleton Coast Park, now a designated wilderness zone. Part of the 1,250-mile (2,000-km) long Namib Desert that fringes the coast of Namibia, the landscape is harsh but beautiful. Rugged brown hills of exposed rock sandblasted by the wind are backed by a distant chain of mountains.

Around the hills stretch broad, gravel-strewn plains with low-lying salt pans and undulating fields of sand. A number of valleys and deep gorges crossing the park mark the courses of rivers from the highland interior. Only one of these, the Kunene, is perennial. Others reach the sea regularly during rainy periods inland, but many disappear before they traverse the desert strip. Though these river courses are dry on the surface for most of the time, water usually lies underground within reach of plant roots, and in places rises to the surface as permanent waterholes.

Sand dunes run inland from the coast. Molded by the wind, they form

On even the brightest day, the Skeleton Coast can be swathed in mist. This treacherous coast claims many victims. Here, a deserted ship lies abandoned to waves of advancing sand dunes.

long ridges, ripples and elegant crescent-shaped dunes known as barchans. Some dunes are anchored around rocks and bushes, while others creep forward as grains of sand are blown over their crests by the wind. Barchans move the most rapidly; some advance up to 50 feet (15 m) across the desert every year.

Given the Skeleton Coast's tropical latitude, the water offshore is surprisingly cool, generally below 59°F (15°C). The chill is meted out by the cold Benguela Current, which flows up from Antarctica and sweeps along the west coast of Africa. The current gives the coastline its most characteristic feature – persistent fog. As warm, moisture-laden air from more westerly parts of the ocean meets the cold air above the Benguela Current, it cools and condenses much of its water vapor. Dense banks of fog usually hang offshore like a giant curtain, hiding the ocean beyond.

For mariners skirting the shore, this sea fog has always been a terrible curse, capable of cutting all visibility in seconds. Temporarily blinded, and on a sea with strong currents and heavy swells harboring hidden reefs, many have run their ships aground.

Even today, with the aid of navigational equipment, ships are still lost now and again along the Skeleton Coast. In times past, castaways from the shipwrecks faced little chance of rescue before thirst and hunger took their toll. It can still take several days for people lost in the desert to be rescued.

Lack of water is the worst problem. Although the sea fog makes the coastal air humid, the layer is too

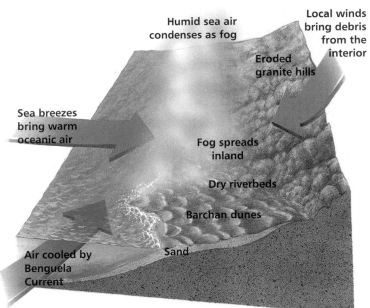

Humid sea air condenses as fog

Local winds bring debris from the interior

Eroded granite hills

Sea breezes bring warm oceanic air

Fog spreads inland

Dry riverbeds

Barchan dunes

Air cooled by Benguela Current

Sand

FOG OVER THE DESERT
Billowing curtains of fog form as warm, damp Atlantic air condenses over the icy Benguela Current, which travels up from the Antarctic Ocean. In early morning, the fog spreads inland, bringing precious drops of moisture to the barren desert.

thin to create rain, and wider climatic patterns mean that the area seldom receives rain clouds. Just a few light showers fall from year to year and these evaporate swiftly in the burning heat of the day.

For castaways, the fruitless humidity of the coastal zone must have seemed all the more frustrating

at night and in the early morning. After sunset, the desert air quickly cools as the heat of the day escapes into the atmosphere. In some places temperatures drop close to freezing. The humid air near the coast becomes damp with condensation like the air offshore and soon fog develops onshore. The desert fog thickens and gradually spreads. By dawn, thick mists cover the desert for about 30 miles (50 km) inland, swirling into

> *"Dense banks of fog usually hang offshore like a giant curtain, hiding the ocean beyond"*

A lone gemsbok roams the Namib Desert in search of food. It rarely drinks but gets all its moisture from the plants it eats. One of the driest of all deserts, the Namib receives no more than 1 inch (25 mm) of rain a year.

The welwitschia plant has long ribbon-shaped leaves which curl over the ground. The millions of tiny pores on each leaf, normally kept shut to conserve moisture, open up when nighttime fog condenses on the plant.

valleys and drifting over hills. Once the sun starts to rise again, the air warms once more, and the land fog gradually disappears.

The extreme aridity of the Skeleton Coast makes the sight of an elephant or even a lion strolling across the desolate landscape all the more incongruous. Yet this seemingly inhospitable desert harbors a remarkable range of wildlife. Animals more at home on the African grasslands manage to survive along the river valleys, relying on the waterholes to slake their thirst when the riverbeds are dry. Elephants, giraffes, zebras and antelopes travel these vegetated corridors, browsing

Cape fur seals come ashore in their thousands at Cape Cross on the Skeleton Coast to give birth to their young and remain there for up to two months until the young are strong enough to brave the sea.

on trees and bushes such as acacia and mopane whose roots draw on underground water reserves.

Lions sometimes lie in wait for prey around watering sites. Troops of chacma baboons have even colonized dry, rocky gorges with tiny erratic water supplies. Normally, these monkeys need to drink daily, but

observations of one group have shown that they have learned to survive without drinking for more than three weeks, if necessary, by reducing activity to a minimum.

Away from the valleys, on the rocky plains and dunes, true desert conditions demand special adaptations in plants and animals. Some plants such as grasses and herbs sprout only when chance rains dampen the desert soil, engaging in a race against time to grow and flower before the water disappears.

But the desert is best known for the plants that exploit its nighttime fogs. The ganna is a common straggly bush which usually traps around it a hummock of windblown sand; its roots then readily soak up fog and dew deposited over the dune. Desert lichens, which are abundant on rocks and gravel, remain dry, brittle and dull until the fogs sweep over them. Then they quickly soften, and their colors become vivid.

Some of the most remarkable of all desert animals live on the dunes. These seemingly barren seas of moving sand have their own hidden ecology, based on mere trickles of incoming nutrients and moisture. Ants, termites and beetles scour the sand for organic material blown from inland and from river valleys. They, in turn, sustain a web of burrowing predators such as scorpions, lizards and snakes.

Some of the dune animals "drink" the morning fog that rolls in from the sea by lying exposed near the dune crest and letting the moisture condense on their bodies. Snakes and lizards lick the precious drops from their skin; the sidewinder viper simply flicks its tongue over its coils. Other animals manage to obtain all the moisture they need from the plants or prey that they eat.

Through ingenious adaptations, wildlife manages to thrive in this most unusual of deserts, but the Skeleton

"These seemingly barren seas of moving sand have their own hidden ecology, based on mere trickles of incoming nutrients and moisture"

Coast has always been a harsh place for people. Few native tribes have ever penetrated the desert proper. European explorers came to the region from the 15th century, but did not linger long.

The 20th century has seen the arrival of prospectors and mining companies looking for diamonds and other minerals. Most of these operations proved uneconomic, and their facilities are now abandoned to the desert. The disfiguring imprint of human intruders who have come and gone – deserted huts, shipwrecks, bones – serve as testament to the wild seclusion of the Skeleton Coast.

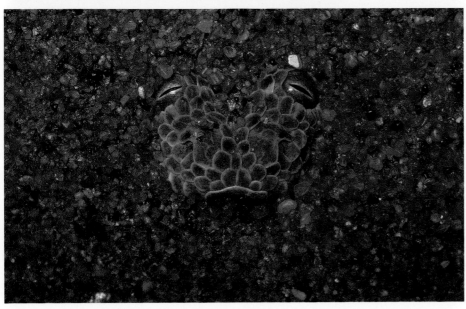

Half-hidden in the sand, a sidewinder viper lies in wait to ambush a passing cricket or web-footed gecko. The snake's mottled coloration helps it stay perfectly camouflaged.

A darkling beetle balances precariously on its head, so that the early morning fog condensing on its body trickles down into its mouth. These precious drops are the only direct source of moisture that the beetle can obtain in the parched desert.

SWEDISH LAPLAND

"An immense wilderness with a harsh and remote beauty, unchanged for thousands of years"

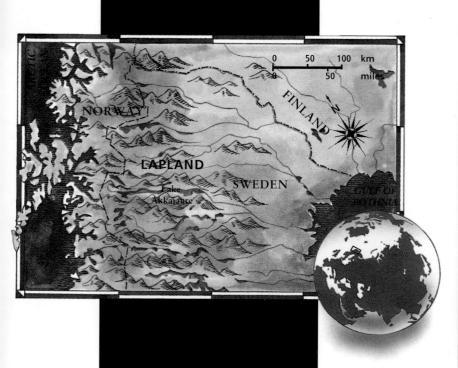

Sweden's vast northernmost county, occupying almost a quarter of the country's entire area, contains some of Europe's wildest scenery. The land sweeps down from the snowfields and craggy peaks of the western mountains in a long, wide slope toward the Gulf of Bothnia to the east. Glaciers, lakes and torrential rivers cut deep parallel grooves across this slope, running from northwest to southeast. As the terrain changes levels, it encompasses snow-covered summits, alpine tundra and deep-carved valleys, descending finally to virgin pine forests, lakes, peat bogs and marshlands.

Much of Lapland is north of the Arctic Circle, and its inhabitants, animal and human alike, have become adapted to a life of long, cold winters and short, cool summers. The few encroachments of the modern world – the mining towns and the tourist centers – are dwarfed by the immensity of the wilderness, and huge areas of Swedish Lapland retain a harsh and remote beauty that has remained unchanged for thousands of years.

The present shapes and forms of the land were sculpted as the great ice sheets of the last ice age retreated northward 10,000 years ago. Relieved of the huge weight of the ice, the crust beneath rose upward. Subterranean forces heaved and tilted the ancient plateau, thrusting the Scandinavian peninsula above the waters. Melting glaciers left behind a rocky detritus which dammed rivers and streams to form the elongated lakes typical of northern Sweden.

Rocky mountain crags look down on vast river systems snaking across the flat plains. In this wild and desolate landscape of Swedish Lapland it is possible to travel for weeks without seeing another human being.

Much of Swedish Lapland is occupied by national parks, which are really wildernesses within a wilderness. They cover a vast area and contain a variety of landscapes rich in plants and animals. Three parks – Stora Sjofallet, Sarek and Padjelanta – are grouped together in the northwest, next to the Norwegian border. Together the parks comprise Europe's largest protected wilderness, with a total area of some 1,292,850 acres (523,200 hectares).

In Stora Sjofallet, steep mountain ranges flank Lake Akkajaure, and primeval forests of Norway spruce and Scots pine in the east of the territory give shelter to large mammals such as bears and elks. Sarek has 90 peaks of more than 5,900 feet (1,800 m) and close to 100 glaciers. Remote and difficult to access, the Sarek land ranges from steep river valleys forested with birch to alpine tundra and rocky peaks. The largest of the parks, Padjelanta, has a wide, rolling mountain character, with lakes bordered by alpine meadows and grassy heathland.

Central Lapland, which includes Muddus National Park, is a region of forests and peat marshes. Its bird sanctuaries shelter more than 100 species, from water birds to golden eagles, and its extensive wetlands are home to otters as well as water shrews and voles. Farther south is the Peljekaise National Park, established to protect the mountain birchwoods that glow with heathers and flowers in spring and summer.

Plant life varies with altitude on the rugged Lapland slopes. Lichens are the only plants that flourish in

The lynx once lived all over Europe, but Swedish Lapland is now one of its few remaining refuges. In these remote mountainous and forested lands, this stealthy hunter preys on hares, birds and even young reindeer.

NORTHERN LIGHTS

The aurora borealis, or northern lights, is a dazzling display of shifting colored light that illuminates the polar night sky.

The phenomenon is caused by solar winds, carrying charged particles, streaming toward Earth from the Sun. The magnetic field of the Earth is shaped like a doughnut, with the holes over the poles. Charged solar particles pour down through these holes and react with the atmosphere's nitrogen and oxygen molecules to produce the colorful lights.

Solar flares periodically boost the density of the particles, leading to particularly spectacular displays. A similar phenomenon occurs in the southern hemisphere.

the upper alpine zone above 4,000 feet (1,200 m). Their success attests to the lack of airborne pollutants in the cold, clear air. The lower alpine zone beneath this level supports a scrub-growth of dwarf willow, and the next zone down, the subalpine, is typified by birch forests. Lower down still is the coniferous zone, with its forests of pine, spruce and fir. Hardy plants such as blue mountain heath, mountain aven and three-leaved rush carpet the ground in the subalpine and lower alpine zones.

The mountain uplands of Lapland are the territories of many birds of prey. Golden eagles, rough-legged buzzards, gyrfalcons and snowy owls all hunt there. Water birds found in the marshes and lakes include lesser white-fronted geese, red-necked phalaropes and whooper swans. The larger wild mammals are becoming increasingly rare in Europe, but the Lapland wildernesses provide refuges for bears, elks, and almost extinct wolves. Wolverines and lynxes still stalk the forests and woods.

Swedish Lapland has until now managed to survive the predations of industry and tourism, and its parks are among the most successful in Europe. The long, dark winters and the harshness of the Arctic terrain have undoubtedly contributed to the preservation of this icy land.

In the highlands of northwest Lapland, well above the Arctic Circle, land over 3,300 feet (1,000 m) is treeless, covered with vast stretches of wind-scoured permanent snow. The few plants that can survive grow close to the ground.

THE SAMI OF LAPLAND

Northern Scandinavia has been home to the Sami, as the Laplanders prefer to be known, since prehistoric times. Their ancestors may have been Stone Age hunters from central Europe who followed the retreating glaciers northward. There are currently about 60,000 Sami in Scandinavia, 17,000 of them in Swedish Lapland.

Some Sami became fishermen, others settled into forest communities, and still others adopted a seasonally nomadic life managing great herds of reindeer. The herders work to an eight-season year.

In April, known as springwinter, the Sami watch over the reindeer as they wander far over the snow-covered plains, digging through the snow for lichens. Spring sees the herds heading for the mountains where the cows give birth to their calves, and in springsummer and summer proper, the herds feed on abundant mountain vegetation. In autumnsummer, the herds are driven down to folds at the foot of the mountains for slaughter, and through autumn the meat is preserved or sold. The reindeer stay close to the woods during autumnwinter, gradually moving farther afield in search of food as winter closes in.

The 3,000 or so Sami who still depend on reindeer speak their own language and organize their own village communities. They survive in their harsh environment as robustly as the reindeer by which they measure their wealth.

In August and February, the Sami hold ceremonial fairs, dressing up in their bright tunics, hats and boots, and buying and selling traditional wares as well as reindeer calves (below). Reindeer are the wealth of a Sami family.

The Sami once had a completely nomadic life-style, moving with their reindeer between winter and summer pastures. Nowadays, most people live in permanent homes while some of the young men do the herding (right).

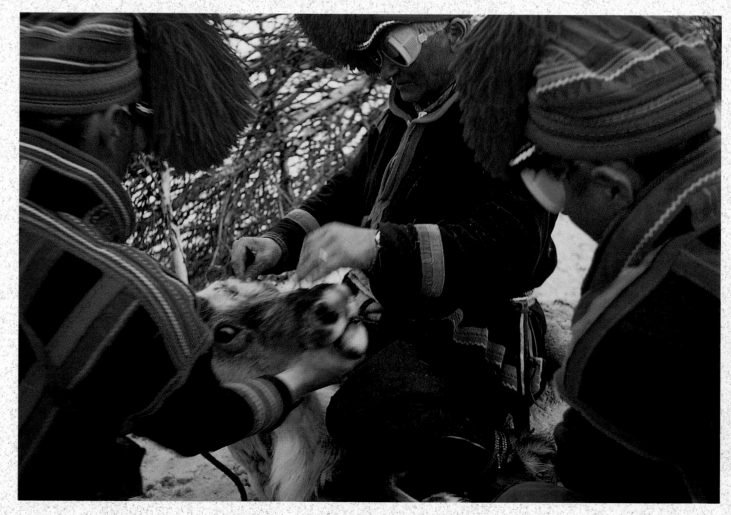

Reindeer are rounded up annually so that herders can mark their calves with earcuts or identifying brand marks. This is necessary since herds spread out during winter movements and can easily become mixed with another family's animals.

THE CAIRNGORMS

"Below the plateaus, cliffs and sweeping moors of the high mountains are primeval pine forests"

Bounded by great river systems, riven by steep-sided glens, and skirted by the remains of ancient forests, the Cairngorms are Britain's largest mountain region. This mighty granite mass in the Scottish Highlands is landscape on a grand scale, dwarfing all human activity.

In this uncompromising terrain, climate and landform conspire to create regions of daunting remoteness. For much of the year, the upper peaks and glacier-scoured slopes are brilliant with snow and ice. People make seasonal incursions to ski, climb, hunt or fish, and then hurry away again, leaving the lochs and rivers to ospreys and otters, the forests and mountains to eagles and deer, the snows to hares and winter-plumaged ptarmigan.

On the broad humps of the high mountain summits, no fewer than six of them more than 4,000 feet (1,200 m) high, bright sunshine can turn into gales and blizzards with frightening swiftness. More snow falls on the Cairngorm plateau than on any other mountain region in Britain; and, in high, shadowed fissures on the summits, snow can persist through to August or even September.

Below the plateaus, cliffs and sweeping moors of the mountains, the Cairngorms contain remnants of the great primeval pine forests that once covered well over 2½ million acres (1 million hectares) of Scotland. For 5,000 years and more, the pine forests formed a tree line up to an altitude of 2,000–2,100 feet (600–640 m) on the Cairngorm plateau.

Scots pines have grown in the Cairngorms since the end of the last ice age. These resinous trees, armored with great slabs of rough bark, flourish in mineral-deficient soils and provide food and shelter for a wide range of mammals, birds and insects.

But from the 16th century onward, the majestic trees were burned to clear land and, later, felled for industrial furnaces. Now only 20,000 acres (8,000 hectares) of natural forest survive, scattered throughout the highlands, particularly on the slopes of the Cairngorms. The Scots pine, which thrives in poor soils and harsh climates, still dominates. The remaining stretches of trees vary from groves of straight-trunked giants almost 100 feet (30 m) high to twisted, wind-stunted trees of no more than 10 feet (3 m) on exposed slopes near the tree-line limit.

These old pine forests of the Cairngorms are found mainly in a curving swath that takes in Glen Feshie in the west, Rothiemurchus in the northwest and the Abernethy Forest in the north. There are forest pockets here that have never been controlled or managed in any way.

Unlike modern pine groves with their rows of uniform, geometrically planted trees, the ancient forest tracts contain open glades where the sun can penetrate the dense top cover. Here, birch, juniper, rowan and aspen grow among the craggy pines, and small plants such as heather and crowberry thrive on the forest floor. Once again, contrasts with commercial plantations are apparent – the ancient forest patches are noisy with birds such as the crossbill, which feeds on the seeds of the Scots pine.

Above the tree line, moorlands and peat bogs sweep up toward the rounded mountaintops. The retreating ice masses and glaciers of the ice age have left their mark on the Cairngorms, where the ice cap was once 2,850 feet (868 m) thick. The ice deepened the river glens, filled lochs with meltwater and terraced the sides of valleys with deposits of gravel. The moving ice also left behind thick layers of clay, sand and gravel which became fertile beds for the great forests.

Ice also carved the amphitheater-like bowls called corries high in the northern and eastern faces of the mountains. Prevailing winds blew snow from the plateau which accumulated in hollows on the mountains. As the climate worsened and temperatures dropped, the compacted snow turned into ice and formed small local glaciers. These gouged the corries out of the mountainside, pulverizing rock and then carrying it away.

The great highland plateau of the Cairngorms consists of several

Thousands of icy winters have shattered these granite summits, already scraped bare by glaciers. Here, on the summit of Cairn Gorm, 4,000 feet (1,200 m) above sea level, few plants or animals can survive.

clusters of high, domelike granite summits, divided by deep glens. Most impressive of these deep divisions is the valley called Lairig Ghru. This great rift lies between the summits of Braeriach, Cairn Einich, Angel Peak and Cairn Toul, to the west, and

VALLEYS SHAPED BY GLACIERS

The Cairngorms owe their rounded forms to the action of glaciers during successive ice ages. The weight of slowly moving ice, aided by the hard debris of fractured rock borne with it,

Before glaciation

V-shaped valley

U-shaped valley

After glaciation

has scoured away the sharp granite peaks and steep-sided valleys, forming rounded summits and U-shaped glacial valleys.

Cairn Gorm and Ben Macdui, highest of the peaks at 4,692 feet (1,430 m), to the east. Beyond Glen Derry, on the easternmost edge of the plateau, lie the high masses of Beinn a' Bhuird and Ben Avon.

South of the entire massif, the glens of Dee, Lui and Quoich run down into the east-flowing Dee which forms the Cairngorms' southern boundary. Rivers surround the whole region. The Feshie in its great forested glen is the western boundary, while the mighty Spey delineates the northwest margin as far north as the

Abernethy Forest. The Avon flows along the east of the plateau, running north from its source on Ben Avon.

Above the forests and woods around the lower slopes of the mountains, the land becomes bare and unprotected, and the plants that survive there have to be able to cope with freezing gales and months of snow cover. Most of these Arctic alpine species grow close to the ground and literally keep a low profile. Some, like alpine mouse ear, have wooly leaves to keep out the cold. Others, such as the stonecrop, have waxy leaves to conserve moisture and withstand severe weather. Ling, or heather, is the most common moorland plant-cover, turning great expanses of the hills up to about 4,000 feet (1,200 m) purple in August. Above the heather is a zone of mosses, deer grass and stiff sedge, and on the highest levels, where the ground surface is a mixture of patchy earth and shattered rock, tough plants

"Retreating ice deepened the river glens, filled lochs with meltwater and terraced the sides of valleys"

such as mat grass and three-leaved rush predominate.

Wildlife thrives in the ancient forest remnants on the lower slopes of the Cairngorms. Herds of red deer feast through the good weather on grass and heather, lichens, mosses and blaeberries. But the swift onset of the long winter drives them down from the open ground to the resin-scented shelter of the trees, where they wreak havoc on pine seedlings.

The appetites of the increasing herds of red deer are causing problems for the Scots pine. Mature trees grow ever larger, but there is no regeneration in the form of seedlings to replace the old trees as they fall. Many of the current Scots pine woods consist of mature trees that were seedlings some 200 years ago, when there were more humans and fewer red deer, and the herds that existed had a greater range of woodland to use in winter. New afforestation lands are usually fenced, which forces the deer increasingly into the unfenced ancient forest areas. Forest management could redress the balance and maintain the survival of both species.

Roe deer do not move around in large herds like the red deer. They stay instead in family groups or travel singly as they roam the slopes browsing on small shrubs such as bilberries. Reindeer became extinct in the 12th century, but had previously lived in the region. In 1952 a domesticated herd of reindeer was brought from Sweden to an area near Aviemore in the Cairngorms. The introduced herd is now living successfully in the wild, feeding mainly on mosses and lichens.

The old forests provide one of the last protective environments for the red squirrel, which has been usurped across most of Britain by the gray squirrel, introduced from North America. Red squirrels are natural inhabitants of coniferous forests,

and pine nuts are a favorite item in their diet. Other increasingly rare mammals still holding out in the Cairngorm pine forests include the wildcat and the pine marten. The otter, too, is found along forest rivers.

The wide and windy summits of the Cairngorms are home to several species of mountain birds. The dotterel, a member of the plover family, winters in north Africa, but migrates to the far north to breed and nest. The birds start to arrive in the Cairngorms in May, though many continue northward to nest well within the Arctic Circle, where they favor sea-level sites. In the Cairngorms, dotterels nest on the high mountain tops, the smaller males sitting on the eggs while the females congregate in groups away from the nests. The snow bunting, like the dotterel, is a rare bird and nests high in rocks and cliffs of the summits. There are maybe 30 pairs nesting in the whole of Scotland.

Another mountain dweller is the ptarmigan, which can survive in the severest of conditions, often roosting in holes in the snow. One nest found in the Cairngorms was at an altitude of 4,400 feet (1,340 m). The ptarmigan has a brownish plumage in the summer, with black barring and white wings. In winter it is all white. It feeds on flowers, shoots, berries and seeds of mountain plants.

But perhaps the most interesting bird found in these mountains is the capercaillie, with its spectacular courtship ritual. This large relative of the grouse became extinct in the 18th century, but was successfully reintroduced into the Cairngorms in the 1830s with Swedish stock.

In the spring, usually before dawn, male capercaillies begin their courting display by emitting a long series of clicks and rattling noises, mixed with hisses and gurgles. Moving into an open glade on the forest floor, a dominant male, resplendent in

colorful blue-green plumage, with red skin patches over his eyes and bristly feathers on his throat, spreads his great fan of a tail, holds his head high, and leaps into the air. Hen capercaillies assemble to watch the display from their roosts, and other males also enter the arena to perform their own, similar displays.

Around 33 pairs of strictly protected ospreys nest in the Scottish Highlands. In the Cairngorms, they nest at Loch Garten and Loch Morlich and, despite the predations of human nest robbers, they continue to thrive. The ospreys overwinter in Africa, migrating north in early spring to reach Scotland in March. One of the great sights of the Cairngorms is that of an osprey beating over a loch on its 5-foot (1.5-m) wingspan, before plunging into the water, wings held high, to seize a fish.

Despite some increase in tourism and winter sports, the Cairngorms, most of which have now been made a National Nature Reserve, are still one of the wildest mountain regions in Europe. Good management could mean that this area, with its awe-inspiring snow-covered summits, ancient forest lands and abundant wildlife, remains the unique and wild place it is at present.

The moorland slopes of this broad pass through the mountains look down on a fertile wooded valley.

The capercaillie nests among the heathers on the forest floor and feeds on berries, leaves and insects (below).

Ospreys build their nests at the top of Scots pines, returning to the same nest year after year (right).

BIALOWIEZA FOREST

"A remnant of the huge primeval forest that once covered much of lowland Europe"

A living link with prehistory, the great Bialowieza Forest is the largest surviving region of primeval mixed-tree forest in Europe. It preserves the conditions which must have existed throughout many European forests two thousand and more years ago. Great oaks, hornbeams and pines create a dense, high ceiling that excludes the sunlight. This dark, shadowed gloom is occasionally relieved by glades where a thinner leaf cover lets through a green light.

Although it covers a total area of 460 sq. miles (1,200 km²) straddling the border between northeastern Poland and the Republic of Belorus, Bialowieza is but a tiny remnant of the huge forested tracts that once covered much of lowland Europe. The strictly protected Bialowieza National Park on the Polish side of the border constitutes some 13,000 acres (5,300 hectares) of the entire forested region. Internationally, the park is thought important enough to be a United Nations Biosphere Reserve and has been designated a World Heritage Site by UNESCO.

The forest is deliberately left untouched to preserve its virgin state. Ancient trees lie where they fall, to be overgrown by climbing plants, mosses and fungi. There is a rich, pungent dampness in the air, and the forest floor is frequently waterlogged. Peat bogs and marshes form open areas among the trees and along the banks of streams and rivers. This park is one of Europe's best inland wetlands.

The survival of the Bialowieza Forest is due to a long tradition of

A remaining fragment of ancient European forest, Bialowieza is a haven for small herds of bison. The animals graze on trees and bushes as well as ground plants. A full-grown bull is a powerful animal and can weigh more than 2,200 pounds (1,000 kg).

> *"King of the forest, the European bison looms out of the shadows like a dark ghost from an earlier age"*

protection going back to at least the 15th century. Generations of rulers – first the dukes of Ruthenia and Lithuania, then the kings of Poland, and finally the tsars of Russia – claimed the forest as their personal hunting grounds. Local inhabitants were forbidden to enter the forest lands to hunt or gather firewood.

By the beginning of the 19th century, the Bialowieza Forest was

Bialowieza contains broad-leaved and *coniferous trees at all stages of growth. Flourishing in the rich earth, saplings shoot up tall and thin as they reach for the sky beyond the forest canopy.*

already the last refuge of the European bison, which had once flourished from the Atlantic coasts of Europe to the China seas. Displaced by forest clearances across the continent, the last few hundred bison survived in the dark and impenetrable glades of Bialowieza – culled only by royal hunting parties – until the revolutions and wars of the early 20th century brought the first real threat for half a millennium to the forest and its inhabitants.

World War I saw the cutting of millions of cubic yards of ancient timber and the decimation of game, including the bison, for meat. The national park was created in 1921,

too late to save the last forest bison, which was shot in the same year. That would have been the end of the story for this primeval beast had there not existed around 60 bison which had been donated to various zoos and private game reserves.

A breeding program, using animals originating from the Bialowieza herd, was begun in 1929, and the first small herd was released back into the wild in 1956. There are now more than 250 wild bison living in the park in the Polish sector of the forest, and a somewhat larger herd in the Belorus sector. The European bison has a humped back, broad head and a tangled mane that includes a beard. Its horns are shorter and broader than those of its American cousin.

The forest is the green lung of a country which has had a bad record of industrial pollution. It is bracketed by the headwaters of the Narew and Lesna rivers, tributaries of the Bug, and the deep, rich soil nourishes immense trees, some of them 500 years old. Oaks and limes soar to 130 feet (40 m) in height, topped only by giant spruces 160 feet (50 m) tall. In total, some 26 species of tree contribute to the forest's great mass. There is also an abundance of flowering plants, with more than 550 species representing over 25 percent of the country's flora.

The birdlife of the forest reflects its physical makeup and the great range of available foods. Five different varieties of woodpecker noisily work the trees for grubs and insects, while the glades, swamps and river meadows are the domain of predators which include eagle and pygmy owls, black kites, goshawks, honey buzzards and several species of eagle. Warblers live among the water plants, and flycatchers and finches feast on the teeming insect life.

Deep in the protected heart of the park, creatures that have disappeared from most of Europe still live and breed and graze and hunt, as they have done for thousands of years, safe from human intervention. Lanky elk wade through the marshes to feed on lush tangles of water foliage. Families of wild boar, guarded by huge old patriarchs, rip up the black soil for roots. Wolves and lynxes slip through the trees like wraiths, seldom seen by visitors. In the rivers, otters prey on the two dozen varieties of fish, while rare European beavers fell saplings and build their dams and lodges. Beavers have succeeded in crossing the border back into the Polish part of the forest after disappearing from there in the early part of this century.

Another great rarity is the stocky tarpan horse, a wild ancestor of the domestic horse, which became completely extinct early this century. Selective breeding of cross-bred zoo specimens has produced an animal which closely resembles the original tarpan, and a small herd is being raised within the park.

But the bison remains the great success story of the Bialowieza Forest. These great beasts are quietly thriving in the depths of this remote reserve. The undisputed kings of the forest, they loom out of the shadows like dark ghosts from an earlier age. The historical record of the European bison goes back to cave drawings of the last ice age. The great triumph of Bialowieza is that it is possible to step into the dark glades of this primeval forest and come face to face with a majestic creature that roamed identical forests in the time of our own Stone Age ancestors.

Ancient stumps and fallen trunks crowd alongside seedlings, saplings and mature trees. There is no artificial forest management.

Once the prey of royal huntsmen, who preserved the forest for their sport, the wild boar now lives peacefully in glades deep in the forest (right).

CÉVENNES NATIONAL PARK

"Granite mountains and limestone plateaus share a wildness and grandeur"

Peripheral zone
Central zone

Causse Sauveterre

Causse Méjean

CÉVENNES NATIONAL PARK

FRANCE

0 12 24 km
0 10 20 miles

The Massif Central is a huge highland territory, covering one-sixth of France's total area, which stretches from the Loire in the north down almost to the Mediterranean in the south. Situated in the extreme southeast of the Massif Central is the Cévennes region, a thinly populated area of granite mountains and limestone plateaus. These two landscapes are completely different in character, but share a wildness and grandeur that make this a natural site for the country's largest national park.

The park, established in 1970, has a central protected zone of 225,894 acres (91,416 ha) surrounded by a peripheral zone. Some 600 villagers still live in the protected zone.

A great fault-line scarp running roughly southwest to northeast divides the Cévennes into two major watersheds and climatic zones. North and west of this line, the rivers flow toward the Atlantic and the climate is oceanic and alpine, with strong winds in the high country and icy winters. South and east of the scarp, the streams feed rivers flowing into the Mediterranean. The eastern slopes enjoy a Mediterranean climate, with hot summers and extremely wet winters and springs.

The mountains of the scarp are forested with pine, chestnut, fir, oak and beech below their rocky summits. Their lower slopes of meadow and heath blaze with wildflowers in spring, but in winter the granite and schist tops are hostile wildernesses of wind, rain and snow. Remote and

Swollen in spring with melted snow and rain, several of the great rivers in the Cévennes have carved deep gorges in the limestone cap. Unpolluted and difficult of access, these river gorges are an ideal refuge for wildlife such as beavers and otters, as well as for rare birds of prey.

wild, these uplands have historically been a refuge for fugitives and guerrilla fighters and are now an ideal sanctuary for wild creatures.

West of the scarp, in the Atlantic watershed, the landscape changes dramatically. Instead of the ancient granite and schist bedrocks of the mountains, a much more recent limestone surface caps the strange plateaus of the Causses region. In fact, this is one great plateau, divided into a series of "causses" by the deeply carved gorges of the rivers flowing west toward the Atlantic Ocean.

The Causses are windswept table-lands with sparse vegetation and no surface water. Winters here are bleak and bitterly cold. The dry summers can be bakingly hot. In the fall torrential rain soaks straight down through the permeable limestone until it reaches a water-resistant layer over which it can run. Over millennia, the mildly acidic rain has eaten out subterranean networks of rivers and caves in the soluble limestone. Within these caves, dripping water has deposited minerals leached from the rock to build fantastic structures of stalactites and stalagmites.

> "Historically a refuge for fugitives and guerrilla fighters, these uplands are now an ideal sanctuary for wild creatures"

The peaks lying along the central scarp of the Cévennes each has its own character. Mount Lozère is in the north of the park, a huge granite mass with a wide mound of a summit where the bare upland grass slopes are interrupted by piles of gigantic granite boulders. The decline of grazing has encouraged heathers and shrubs to begin the long task of recolonizing the dense grass mat of the upper slopes. Lozère is the region's highest peak at 5,574 feet (1,699 m), with stunning views over the entire Cévennes hill country to the south.

Mount Bouges lies just southwest of Lozère. The north-facing slopes of Bouges are granite, and well forested between 1,640 and 4,600 feet (500 and 1,400 m). Evergreens are the predominant trees, interspersed with beech forests. The southern slopes of the mountain are of schist, and the terrain is typical of the Midi, with steep, sun-bleached ravines, and groves of chestnut and scrub oak.

Mount Aigoual means "the watery mountain" and lives up to its name. Its annual rainfall of 89 inches (2,250 mm) makes it the wettest place in France. Occupying the southwestern end of the scarp, Aigoual bears the brunt of the rains released as the Atlantic and Mediterranean air currents converge. The climate here is Mediterranean on the southern side, with steep valleys and chestnut groves. The long, dry summers are in marked contrast to the deluges of winter.

To the north, the mountain is rich in conifers such as Scots and Austrian pine, Norway spruce and larch. The slopes on Aigoual's northern side are far less steep than those on the south and extend to meet the limestone plateaus of the Causses.

The limestone cap of the Causses was laid down when the region was covered by sea in the Jurassic age. Nowadays, the only water to be found is deep in underground caves and channels, or in the rivers that swirl through the deep gorges to divide the limestone like the slices of a cake. In the park, the Causse Sauveterre lies

The exposed plateaus and moorlands of the Cévennes region suffer severely cold winters. The plants which do manage to survive among these gale-blasted mountainous outcrops are hardy and stunted.

between the rivers Lot and Tarn, and the Causse Méjean lies between the Tarn and the Jonte. The River Tarn runs through a spectacular series of steep-sided gorges towering above the rushing river for more than 30 miles (50 km) on its way west. Despite harsh extremes of winter and summer weather, the Causses put on tremendous displays of wildflowers in the spring, including wild tulips, pasque flowers and dwarf daffodils.

The centuries-long erosion of wild habitat in the Cévennes gradually led to the extinction of some wildlife species, but there has been an effort over the second half of the 20th century to reestablish some of them. The mountains and gorges are ideal environments for some of the larger soaring birds, and both griffon vultures and bearded vultures (also known as lammergeiers) have been reintroduced into the park.

Other reintroductions include red deer, roe deer and European beaver. By the beginning of the 20th century, the beaver had become almost extinct in France, but from a small population surviving on stretches of the Rhône, it was brought back to several regions, including streams in the Mediterranean watershed of the Cévennes. In 1977 the beaver was introduced to waters on the other side of the scarp.

The steep,.rocky terrains of the Cévennes in all their diversity have preserved it from the excesses of commercial exploitation and tourism, It remains an impressive wilderness as well as a superb wildlife reserve, with a wide spectrum of upland habitats and landscapes.

***Snow and rain soaking through the** porous surface of the Causses have created cathedral-like caverns. Water seeping into these caves dissolves calcium and mineral salts, redepositing them as stalagmites and stalactites.*

THE LAMMERGEIER

The magnificent lammergeier lives in remote mountainous areas of Europe, Africa and Asia. Now rare in Europe, lammergeiers had disappeared from the Cévennes, but have been successfully reintroduced into the park, where they are strictly protected.

Lammergeiers, also known as bearded vultures because of the bristles on each side of the beak, soar on their long, narrow wings over the precipitous river gorges of the region. Like most vultures, they do not usually kill prey, but feed on carrion – the remains of dead animals. They often shatter the bones of carcasses to get at the marrow they contain by dropping them from a great height onto rocks below. Lammergeiers also kill small rodents and birds.

COTO DOÑANA

"Winter rains flood the great plain to create marshes stretching as far as the eye can see"

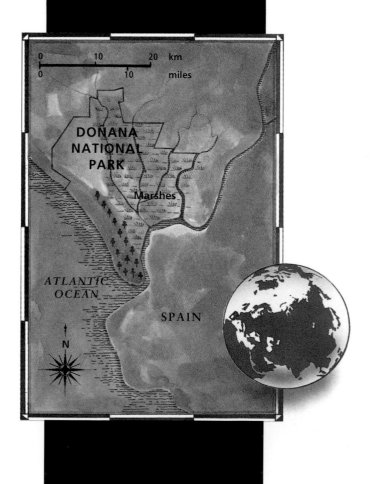

A magnificent wilderness of coastal dunes, marshes, scrubland and sandy heath, the Doñana National Park, in the province of Huelva, is Spain's largest protected park. The region, covering more than 292 sq. miles (757 km²), was once the delta of the Guadalquivir River, which forms the park's southeastern border. A constantly growing sandbar, sculpted into moving dunes by the prevailing

The tranquil and undisturbed wetlands of Coto Doñana provide seashore, dune, flood plain and forest habitats for a wide range of birds and reptiles.

winds, has sealed off most of the delta from the sea, creating a unique barrier. This sandbar contains the vast marshes which make the Coto Doñana park perhaps the most important wetland site in Europe.

When the autumn and winter rains flood the great flat plain to a uniform depth of some 12–24 inches (30–60 cm), the marshes, or *marismas,* of the Coto Doñana stretch as far as the eye can see. Small islands known as *vetas* stay dry throughout the year, providing nest sites for many birds.

But at the height of summer, the marshes dry out in the blistering sun, reminding visitors that this corner of Spain is Europe's closest point to the African continent. What in the wet season was an area of hundreds of square miles of shallows is reduced to three long, canal-like, permanent bodies of water – once delta arms of the Guadalquivir – a few permanent lagoons, some stagnant pools and vast expanses of rapidly baking mud. The reduced waters in the pools can become so contaminated by concentrations of bacteria that they are lethal to water birds.

The great dunes of the Doñana sandbar vary in width between 330 feet (100 m) and 3,300 feet (1,000 m) across the base. Little by little, these dunes are migrating inland as their peaks and ridges are broken down by

winds from the sea. Each year they can move as much as 20 feet (6 m), and anything in their path is buried. Even the sturdy stone pines in the woodlands that border the dunes sometimes fall victim to the sand. The dunes form a barrier all around the tree, eventually burying it completely.

Doñana's proximity to the narrow Strait of Gibraltar makes it a vital gathering, resting and feeding ground for birds which need to take the shortest possible migratory route between Africa and the European mainland. Large-winged birds such as kites and eagles are not able to make the long ocean crossings undertaken by some migrants – they need to rest during journeys by gliding. To do this, they require the thermal currents that are created only over land to carry them high enough to glide without losing too much altitude.

The Doñana National Park is, therefore, particularly rich in large birds of prey, some of them extremely rare. The most celebrated – and the most rare – is the imperial eagle. These majestic birds, with wingspans of up to 82 inches (210 cm), pair for life, and return to the same nest each year. There are at least 15 pairs of imperial eagles in the park – one-third of the entire Spanish population. Other birds of prey include booted eagles, short-toed eagles, red, black and black-winged kites, Egyptian vultures and griffon vultures.

Practically all known species of European waterfowl appear in Doñana during the winter and the main migration periods. The rich wetlands have become even more important to these visitors as other marshy regions of the Spanish coast have been drained for agricultural purposes. Rare crested coots breed in the park, and flocks of flamingos come in winter to feast on the shrimps of the brackish lagoons.

Doñana is also renowned for its populations of herons, egrets and spoonbills. The cork oaks, which thrive on the acid soils of Andalusia, are favorite nesting trees of these social birds, which gather in noisy colonies in the tree tops.

The successive wet and dry environments of Doñana suit reptiles as well as birds, and the park is home to the spur-thighed tortoise and two species of terrapin, as well as many snakes and lizards. Short-toed eagles, armed with tough head feathers to ward off venomous bites, feed on the snakes.

The rare Iberian imperial eagle has one of its few remaining refuges in the Coto Doñana, where it nests on cliff ledges and at the tops of tall trees. It soars high on its broad wings in search of rabbits, which are its main food.

An abundance of predators is always a good sign in any wild environment, since it indicates flourishing populations of prey creatures. Doñana is one of the final refuges of the Iberian lynx. More

LAND OF SHIFTING DUNES

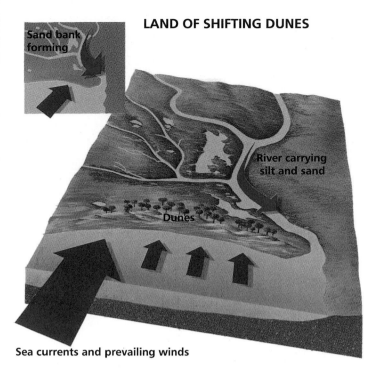

Sea currents and prevailing winds

Coto Doñana is an ancient delta, enclosed by deposits of silt and sand. Over centuries the river has washed down huge quantities of soil which clogged all but the main channel and built up a barrier across the delta mouth. Sea currents and prevailing winds have created a beach along the barrier and piled the sand into dunes.

As the shifting dunes move steadily inland, they surround and bury stone pines growing in their path (above right).

Brackish lagoons and mud flats are ideal homes for noisy colonies of flamingos (right).

than 25 pairs of these stealthy hunters roam the park scrubland, preying on rabbits, water birds and deer calves. Wild cats, polecats, otters, foxes and Egyptian mongooses are other mammalian predators in Doñana.

Up until the early 1960s, Coto Doñana was entirely without access roads. The huge wild area, containing heaths and woodlands flanking flood land or muddy marshes according to the season, was unappealing to impatient developers and much too expensive to turn into agricultural land. In 1969 the region became a national park so that this unique environment and its inhabitants would be protected.

Although rich in wildlife and natural terrains, Doñana is a wilderness under a number of serious threats. Wetlands are notoriously susceptible to disruption, having a complex and delicate balance of seasonal water supply, plants and animals. Water flowing southward into the park from the agricultural developments to the north is vulnerable both to diversion and to pollution. In 1986, some 30,000 birds died in the park as a result of an inflow of agricultural pesticides.

Other threats come from the growing pressures of tourism around the park. The dunes at the nearby resort of Matalascanas have been destroyed by people walking on them, and straying wildlife, including rare lynxes, faces a greater risk of being run over by increased local traffic. Summer tourism is also a drain on water resources, just at the time of year when drought is most likely. Doñana is a superb reserve of international importance, but it is also a good example of the warning that erecting a fence around a site is no guarantee of its security and survival.

THE EMPTY QUARTER

"Lying at the heart of southern Arabia is the greatest continuous sand desert in the world"

The name could hardly be more expressive. Rub al-Khali, the "Empty Quarter" of Arabia, is the very essence of wilderness. In a part of the world that cradled early civilizations, where the paths of adventurers, traders and wandering peoples have crossed for thousands of years, this region has remained forbidding and

The vast sand sea of the Empty Quarter is so difficult for human travelers to penetrate that, even into this century, much of it remains unexplored.

unknown, an empty desert of towering sands and blistering heat. Only scattered Bedouin tribes dared to penetrate this land, but even they could not linger there during the intolerably hot months of summer.

Lying at the heart of southern Arabia, the Rub al-Khali is the greatest continuous sand desert in the world. Up to 300 miles (500 km)

wide, it stretches for some 700 miles (1,100 km) across Saudi Arabia east into Oman. Everywhere it is dry, empty and devoid of soil. Across most of the desert, the sands are sculpted into giant, ancient dunes. Between these permanent features, smaller, more mobile dunes may appear, along with the occasional small salt flat nestling in a depression

in which, every few years, a puddle of rainfall briefly collects. The area, however, has not always been so arid. Large salt flats near the margins of the Rub al-Khali show where lakes once existed, and probing farther back in geological time, it is clear that conditions here have varied greatly.

Pure sand deserts such as the Rub al-Khali are not as common as is often thought. For so much sand to accumulate in one place there must have been a copious source. The bulk of the sand in the Empty Quarter probably came from volcanic highlands to the west and south. Over the millennia, as the elements gradually weathered away the exposed crystalline rock, fragments were washed downstream onto the lowlands. Winds gathered up the smaller particles and swept them farther into the interior.

Another major source in the past would have been large patches of seabed left high and dry by fluctuations in the level of the ocean to the south and east. Winds driving over the exposed sands would have transported more grains to the accumulations inland.

Those winds blowing in the same prevailing pattern for thousands of

Occasional bouts of heavy rain can bring the desert into bloom. Plants spring to life (above) and, by tapping underground moisture, may stay green for months.

As desert sands absorb the intense midday heat, the surface of the dunes becomes scalding. Desert creatures must save their activity for the cooler hours.

years shaped the sands into the massive dunes that persist today. The formations are most regular in the west, where powerful winds have created a corrugated landscape of parallel ridges and troughs. Toward the center of the Rub al-Khali the pattern becomes less well defined, with more dune ridges running crosswise, while in parts of the east, complicated wind regimes have erected and sculpted huge dunes up to 1,000 feet (330 m) high.

Winds in the Rub al-Khali arm the air with abrasive sand. Daytime temperatures soar above 122°F (50°C) in summer, and no rain may fall on a patch of sand for years. Conversely, warmth escapes into the cloudless sky so fast that, after dark, a winter's night may chill below freezing.

But even this land, empty of resident humans, is far from lifeless.

Plants appear here and there, not just as solitary clumps, but thick enough in places to create patches of greenery. At the foot of a dune, there may be small shrubs, saltbushes, tussocks of sedge and short-lived herbs. And, where there are plants, a few animals, such as beetles, butterflies, spiders, scorpions, lizards, larks, gerbils and hares, manage to survive. Still rarer are bustards, sand cats, foxes and gazelles.

One of the principal problems of survival in the desert for any organism is the permanent shortage of water. The ways in which plants and animals cope with drought reveal the sheer adaptability of desert life. Plants have two main strategies. Some are many times more extensive underground than above. They have deep probing roots to tap droplets of moisture hidden deep in the sand, but tiny leaves and stems to minimize water loss through evaporation.

Other plants lie dormant during the long droughts, either as seeds in the sand or as seemingly lifeless, shriveled stems. When a good, steady bout of rain falls, they quickly germinate or burst into greenery, their wide-spreading shallow roots making full use of the moisture before it evaporates. Within a few weeks,

they will have flowered and seeded anew. Shallow-rooted plants also benefit from morning dew.

Some desert insects live in much the same way, lying hidden in the sand in a watertight dormant phase, until stimulated into emergence by heavy rain and fresh plant growth. Larger animals keep active all year and have a remarkable ability to survive without drinking. Herbivores such as rhim gazelle gain enough moisture by eating leaves and roots and by gleaning dew from plants.

Carnivores get most of their water from the body fluids of prey – hoopoe larks, for example, from beetles, sand cats from gerbils and lizards.

Such creatures can get by with so little partly because they are able to conserve moisture. Desert-adapted mammals, for instance, tend to perspire less, urinate less and lose less water vapor when they exhale. For animals and plants simply to survive in the desolate world of the Empty Quarter demands much of nature's evolutionary ingenuity.

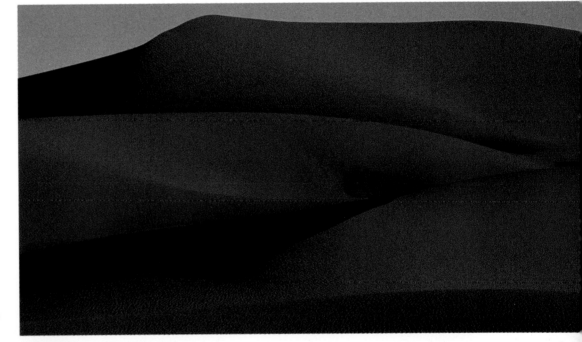

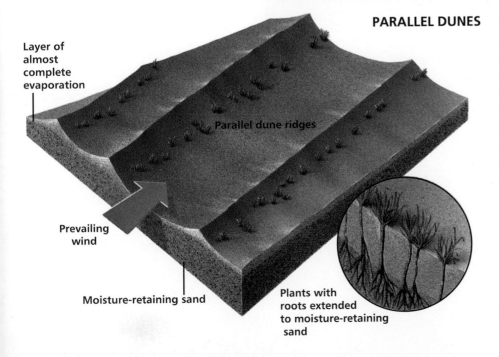

PARALLEL DUNES

Layer of almost complete evaporation

Parallel dune ridges

Prevailing wind

Moisture-retaining sand

Plants with roots extended to moisture-retaining sand

Occasional rainfall percolates beyond the reach of evaporation, creating a layer of slightly moister sand. At the bases of parallel dunes, where this layer is closer to the surface, hardy plants with deep roots can tap just enough water to survive.

The big dunes of the Rub al-Khali are permanent desert landmarks (above). Though winds constantly pattern their surfaces and skim grains across their crests, the underlying shape and position of these dunes changes little over the years.

THE BEDOUIN OF RUB AL-KHALI

If any people can be said to make the Rub al-Khali their home, it is the Bedouin. These hardy nomadic people, for whom the Arabian interior has been a heartland for some 4,000 years, remain the only humans with sufficient knowledge of the dunes and of the perils of the desert to guide travelers in the Empty Quarter.

Bedouin tribes have long lived around the fringes of the zone. They venture into its heart in winter – when temperatures are more bearable – to lead their herds to the patches of scant but fresh pasture, which spring up after the occasional rains. Though the number of Bedouin who lead nomadic lives has been dwindling for decades, a few families, with little but their tents, cooking utensils and livestock, still do so in the Empty Quarter. Clothing consists of long robes and large headcloths to protect them from excessive exposure to the strong sunshine.

The Bedouin move according to the needs of their animals for pasture and for sources of water. In turn, the herds – principally camels, but also sheep and goats – provide them with nourishment in the form of milk products such as curds, buttermilk and cheese, and with meat.

Nomadism in such an open environment requires a flexible form of shelter, and the Bedouin have the perfect solution in their long, low tents of woven camel or goat hair. Easily rolled up for transportation by camel, the tent is secured with ropes over a series of upright poles. The sides can be lifted up by day to improve ventilation or lowered for privacy, warmth at night, or for protection against sandstorms. The design of the tent also reflects the hospitality customary in Bedouin culture. It has an open area to one side for socializing, receiving guests and holding tribal meetings.

Bedouin men may set out on foot across the sands in the morning to tend their camels, not returning to the shade of their camp until evening (left).

Portable, shady and flexible in construction, the typical Bedouin tent (above) is perfectly suited to a nomadic lifestyle as well as to the harsh environment.

Containers and cooking utensils are among the few material possessions of those Bedouin who still lead traditional nomadic lives (right).

Camels have long been the mainstay of Bedouin tribes, a reliance reflected in the care with which the men tend their herds (below).

THE GOBI DESERT

"One of the world's largest arid zones, the Gobi suffers blistering heat in summer, yet bitter cold in winter"

Enormous empty tracts of barren rocks, stone-littered ground and shifting sands, alternating with rolling hills and salt-encrusted depressions, characterize the lonely wilderness of the Gobi Desert. The harsh grandeur of this vast inland region, divided between Mongolia and northern

Set in the heart of the Asian continent, the bleak Gobi Desert comprises a wilderness of sand, gravel and pebble-strewn plains.

China, is accentuated not just by the aridity of its climate but also by seasonal extremes of temperature that bring blistering heat in summer, yet bitter cold in winter.

Long regarded as a forbidding natural barrier between the Russian steppes and the Chinese heartland, the Gobi Desert is one of the world's largest arid zones. From the east, where it blends in to dry grasslands, to the west, where it connects with other Asian deserts, the Gobi stretches at least 1,000 miles (1,600 km).

With an altitude varying between 2,300 and 5,000 feet (700 and 1,500 m), the Gobi forms a vast upland basin, with no river connection to the outside world. Streams flow into the Gobi from a series of high, fringing mountain ranges, but none flows out. All either dry up in the desert heat, drain into hollows filled by salty lakes or marshes, or disappear underground through coarse sand and gravel.

Subterranean water, beyond the reach of evaporation, is widespread in the Gobi. The ground surface, by contrast, is almost completely dry with little if any soil. In places, the surface is one of smooth, polished rock. Elsewhere, winds have stripped the sand away, but left behind larger stones so that the ground is dotted with pebbles.

Across large areas of the Gobi, in both rocky and sandy landscapes,

plant life has scant hold. Usually it is restricted to a few grasses and herbs and stunted drought-resistant shrubs like saxaul and wormwood. Salt-tolerant plants can form more dense cover around the depressions where rainwater collects, and tamarisks may appear wherever their deep roots can tap subterranean water. Around some of the few permanent

waters in the desert, oases of poplars and Siberian elms grow.

But the overwhelming impression proffered by the Gobi landscape is one of barren hills and dry plains. Few parts of the desert receive more than 8 inches (200 mm) of rain a year, and that comes in isolated showers. The dryness is mainly a result of the huge distance that prevailing winds from the west and north have to travel overland before they reach the Gobi. After crossing western Eurasia, they have hardly any moisture left to release.

With cloudless skies presenting little barrier to the sun's rays, summer daytime temperatures soar in the Gobi. At midday it can reach 113°F (45°C) in the shade, and rocks on the ground surface may be scalding to touch. The intense heat makes the desert an unwelcoming place and yet, like all deserts, the Gobi is not as empty of life as it may at first seem. Seed-hoarding rodents – mainly gerbils and jerboas – occur wherever there is some vegetation. Jerboas hide underground in their burrows during the heat of the day, waiting to emerge in the cool of night. Geckos – small, nocturnal lizards – employ the same strategy, though they venture out to hunt insects such as beetles.

Winter brings strikingly different conditions. Because of its high

Slopes on the mountain fringes of the Gobi, where rainfall is slightly higher, can support marginally more plants and wildlife than the desert itself. Here a large flock of blue hill pigeons, at home in rugged terrain, takes flight across a steep valley.

The Mongolian gerbil is one of the most abundant animals in the Gobi Desert (left). It forages for seeds, roots and green plant matter.

The Gobi is one of the world's cold deserts. Winter snows can linger in sheltered spots until spring.

latitude, the Gobi is extremely cold in winter. Strong, chilly winds from the north sweep across the desert, bringing temperatures down at times to -40°F (-40°C). They also whip up violent sandstorms and showers of snow. For months the temperature remains below 32°F (0°C).

Many of the larger mammals that eke a living from the desert have to

> "Chilly winds whip up violent sandstorms and sometimes bring showers of snow"

rely on special insulation provided by body fat and thick hair if they are to survive the cold. Pallas's cat has not only long fur, but also a short nose and short legs to minimize the surface area exposed to the elements. The Bactrian camel – the two-humped central Asian species which survives in the wild only in the Gobi Desert – grows an especially long, shaggy coat ready for each winter. Such an adaptation is essential, since an animal of the size of the camel has no hope of finding shelter in the open desert environment.

Several desert birds, including Pallas's sandgrouse, desert wheatears and short-toed larks, opt for migration to avoid the worst weather. Many of them head to arid parts of southern Asia or even north Africa, places where the climatic variations are not quite as extreme as in the extraordinary hot, cold and ever-dry world of the Gobi Desert.

Lake Baikal

"To refer to Baikal as a lake at all is to insult what is patently a great and powerful inland sea"

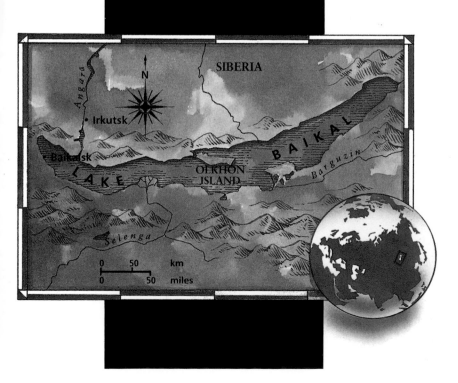

One of Russia's supreme natural treasures, Lake Baikal is set in the vast, sparsely peopled heart of Asia, near the edge of the Mongolian steppe. Stunning mountain backdrops viewed through crystal clear air create a canvas on which the waters paint their changing scenes, from stormy gray, through tranquil blue to the silvery-white glint of winter ice.

Baikal is a lake quite unlike any other: bigger, deeper and older. It fills an enormous trench in the Earth's

surface, and its deepest portion yet fathomed lies 5,370 feet (1,637 m) beneath the surface. Such depth, given its 395-mile (635-km) length and average width of around 30 miles (50 km), means that it holds more water than any other lake in the world – more than the five Great Lakes combined.

The lake has also been in existence for an extraordinary stretch of time. Estimates suggest that it is 25 million years old – few other lakes on Earth can be dated back more than 20,000 years. During those millions of years, flora and fauna isolated in the lake have had time to evolve myriad new forms unknown elsewhere. Baikal has about 2,500 recorded species of animals and plants, an incomparable 1,500 of which are unique to the lake.

Baikal remains cloaked in mystery and there is still much to learn of its natural history. What is known in no way diminishes the sense of wonder with which it has always been regarded. To the Buryat people who have long lived around Baikal, and

A watery wilderness, Siberia's Lake Baikal is the deepest of all lakes and holds one-fifth of all the world's fresh water. The rugged terrain encircling this great inland sea ranges from windswept mountain peaks to vast swaths of dense coniferous forest.

A RIFT IN THE EARTH'S CRUST

Lake Baikal lies in a vast rift valley, created when two faults in the Earth's crust moved apart. The rift itself is more than 5½ miles (9 km) deep, much of it filled with sediment. Life can live at greater depths here than in other lakes. This is because the lake's water is constantly circulating, carrying oxygen to all parts of it.

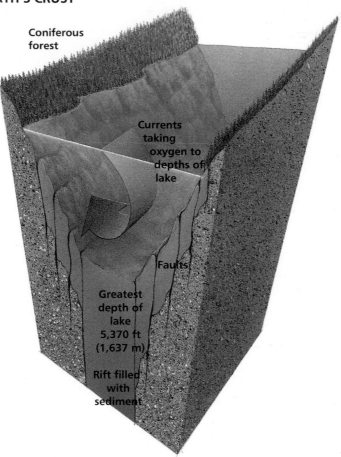

Coniferous forest

Currents taking oxygen to depths of lake

Faults

Greatest depth of lake 5,370 ft (1,637 m)

Rift filled with sediment

***On the densely forested eastern** shores of Baikal, a nature reserve has been established to protect such rare Siberian creatures as the sable, a silky-furred relative of the weasel.*

to generations of Russian settlers in villages and towns along its wild shores, the lake is considered sacred. Indeed, to refer to Baikal as a lake at all is to insult what is patently a great and powerful inland sea.

Baikal's origins lie in the intense geological forces that stress and dislocate sections of the Earth's continental crust, fracturing fault lines and thrusting up mountain blocks. The lake occupies a deep rift valley, a huge chasm of sunken land flanked by rising blocks. To the east and to the north, the wall is especially high, forming snowbound mountain ranges rising 6,500 feet (2,000 m) and more above the lakeside.

Even the water that fills the Baikal chasm is special. More than 300 rivers drain into the lake, bringing with them heavy loads of sediment, along with all sorts of organic detritus and microorganisms. Yet the water that

drains from the lake's only outlet, the Angara River, is almost as clear as distilled water. Water lingers in the deep trough for so long – it has been suggested that incoming water may circulate through the lake for 400 years before it finally flushes out through the Angara – that the finest of sediments settle out, and lake creatures have plenty of opportunity to filter out edible particles and microbes. The lake water is so pure that local people even use it to fill their car batteries.

Winter is hard in Siberia, with temperatures dropping far below zero. Average winter temperatures around Baikal drop to -4°F (-20°C), and the surface of the lake freeezes over for months on end. (A short depth below the surface, the water remains constantly cold, but fluid whatever the season.) As soon as the Siberian summer finally completes the thaw, fishing and cargo boats take

readily to the water. But all boat crews remain wary. Fierce winds howling down river valleys can churn up the water within minutes, the strongest creating treacherous waves. In summer dense banks of fog often develop over the water, obscuring all.

Strong winds and the sinking of chilled surface water each winter may help to explain the extraordinary degree of water circulation within Baikal. Currents take fresh oxygen to the very depths of the lake, sustaining life at all levels. In Africa's tropical Lake Tanganyika, the second-deepest lake in the world, limited circulation makes the lake virtually lifeless below 650 feet (200 m).

Far below the surface of Baikal, by contrast, organisms can survive, as long as they are adapted to the blackness and intense water pressure. More than 50 species of fish live in Lake Baikal, including the curious golomyanka, with its almost transparent body; the sturgeon; the salmon-like omul – an important food fish for the region – and no fewer than 22 species of bullheads.

Shrimp and huge columnar sponges abound in the lake. Both help keep the lake water clean by feeding on organic debris, but the creatures most responsible for Baikal's purity are the legions of tiny epischuras which filter out bacteria and algae. These minute crustaceans crowd in the water at densities of as much as 400,000 per sq. foot (3 million/m^2).

Baikal has yet another zoological surprise in store. It is the home of the Baikal seal, the world's only freshwater seal species. Around 70,000 of these mammals live in the lake, feeding on fish and regularly hauling themselves out to rest on tranquil shores and islands. Quite how seals came to reside so far inland is uncertain, but it has been suggested that their ancestors may have migrated from the Arctic up one of the great Siberian rivers during one of the ice ages.

The Baikal seal may be the only mammal living in the lake, but many others dwell in the wild, forested slopes that ring its shores. These forests, dominated around Baikal by larch, birch and pine, are part of the taiga – the vast belt of coniferous forest that stretches across Siberia. This is the realm of sables, brown bears, lynx and elks, of grouse and owls, of crowlike nutcrackers and the black woodpeckers.

To protect such creatures, several reserves and two national parks have been designated on Baikal's shores. The need to protect the lake's natural riches has become ever more pressing as increasing numbers of people have

"Fierce winds can churn up the water within minutes, creating towering treacherous waves"

settled in the region and industrial development has expanded in a way at odds with Baikal's fragile purity. By the 1950s and 1960s, pressure from fishing led to a collapse in the stocks of omul. Logging on the forested slopes increased siltation, and the practice of floating rafts of logs has resulted in many sinking, clogging rivers and bays.

Worse still, effluent from factories and urban waste, some of it carried to the lake from upstream, has brought

A FOREST OF SPONGES

Baikal water is the purest in the world, and the swaying forest of bright green, freshwater sponges which live on the lake bed at a depth of more than 650 feet (200 m) helps guarantee this purity. As they feed, these sedentary creatures filter tiny organisms such as bacteria and algae from the water, thus preventing it from becoming laden with clogging green slime.

Legions of freshwater shrimp living on the sponges also eat such debris and even clean the filtering pores of the sponges, so increasing their efficiency.

toxic pollution, killing some sections of the lake bed and hampering the recovery of fish stocks. Many polluting sources are to blame, but most notorious is the Baikalsk cellulose plant, which has discharged thousands upon thousands of tons of effluent into the air and water.

Years of protest against the plant and of concern for Baikal's ecosystem in general have at least mitigated some of the damage. Effluents are now treated more thoroughly at Baikalsk than when the plant first opened, and fishing and logging are more closely regulated than before. But most Baikal people feel that much more change is needed.

A lake ecosystem founded on the purity of water that lingers in the basin for hundreds of years simply has no room for industries that pollute or despoil. As local people campaign for their environment, scientists continue to probe for better knowledge of this remarkable watery wilderness. Lake Baikal is still a magnificent jewel of nature, but it is one whose luster could so easily fade.

In winter, when temperatures plummet, Baikal freezes over. But Baikal water is so pure that, although the ice can be more than 3 feet (1 m) thick, it is almost as clear as glass.

THE FISHERMAN
OF BAIKAL

For the people of Baikal, the lake is more than a vast expanse of water. It is a friend, a provider, even a symbol of spirituality. People's lives are linked to the lake, and they depend on its bounty.

Fish are a vital food source in this harsh Siberian environment, and fishing is a traditional way of life. There are more than 50 sorts of fish in Baikal, but perch, grayling and the salmonlike omul are the bulk of the catch. The omul lives only in Baikal and is a great local delicacy, at its best roasted over an open fire on the lake shores.

But Baikal is volatile and can suddenly change from friend to foe when the mighty storms strike. Winds, such as the *sarma*, can whip up in an instant, lashing the lake's calm waters into a violent frenzy and making the fishermen fear for their lives.

Fishing continues even when Lake Baikal is frozen over in the depths of winter. Fishermen go out onto the ice and cast their nets through holes broken in the surface. Here, some of the day's catch is being cooked over a fire lit on the frozen shores.

After a life spent fishing in Baikal's pure waters, this retired fisherman (above) tends his cottage on the shores of the lake that he loves . Over his door is a painting of this "Pearl of Siberia," as the locals call the lake – a constant reminder of its beauty and power.

Baikal's deep waters are bitingly cold, and even at the height of summer there are no easy pickings (left). Much of the summer catch is preserved by smoking or salting and is then stored for use in the winter months, when food can be extremely hard to come by.

THE
PAMIRS

"Lines of towering peaks soar above deep ravines carved by torrential rivers"

Startling scenic contrasts characterize the mountains of the Pamirs, at the heart of central Asia. In this remote landscape, the scenes juxtaposed by nature are often unexpected. A towering, jagged crest may rise from gentle slopes that lead into a broad, shallow lake; the icy glint of a gray-white glacier may reach out across a sandy brown plain devoid of moisture.

The Pamirs form a great uplifted rectangle of highland at least 155 miles (250 km) across, occupying the eastern half of the Republic of Tadzhikistan. Apart from a series of long, narrow valleys in the west, almost all the land within the Pamirs is more than 10,000 feet (3,000 m) above sea level.

A complex of spectacular ridges – some running west to east, others north to south – raise their peaks much higher still. The tallest, Communism Peak, stands at a mighty 24,590 feet (7,495 m). Like many of the mountain chains, it is mantled with permanent ice descending into valleys as glacier tongues. One of them, the immense Fedchenko glacier, stretches for 48 miles (77 km).

The Pamirs, impressive enough in themselves, stand at the meeting point of several great central Asian ranges: the Himalayas, the Karakorams, the Kunlun, the Tien Shan, the Gissaro-Alai and the Hindu Kush. Where the chains meet, so do their fauna and flora, mixing to give the Pamirs a cross section of life from different quarters. Nevertheless, the wild community is sparse. Cold, dry and stony, the Pamirs present a testing

From the towering summit of Communism Peak, the view stretches across countless crests and ridges in the western Pamirs. Thick yearly deposits of snow on these lofty mountains help to nourish permanent ice caps and as many as 3,000 glaciers.

habitat in which only the most hardy plants and animals can survive.

The tremendous height of the present-day Pamirs is testament to the power of geological movements that have warped and dislocated the entire central Asian mountain system into being. On the evidence of frequent earthquakes in the Pamir region, the mountain-building pressures may still be at work. But the process of uplift is not the only one in action. Erosion, which always works at a faster rate in mountain areas, acts in opposition, constantly wearing down the peaks. Rainfall and frost weather the uplifted rocks; running water and glacier ice, armed with rock fragments, scour slopes and gouge out valleys.

Erosion, on the whole, appears to have been most intense in the western Pamirs. This, along with climatic variations, has given the western and eastern sections of the range strikingly different characters, with a marked contrast in terrain. In the west the valleys tumbling down into the lowlands of Tadzhikistan tend to be

deep and narrow, the high ridges more dramatic in classic Himalayan style. In the east, by unexpected contrast, the relief is one of a high, rolling enclosed plateau, with broad valleys and gentler summits. The east is both drier and colder.

Along the steep, straight valleys of the west, the Pamirs are arguably at their most spectacular. Deep ravines carved by torrential rivers line the valleys that may be 6,500 feet (2,000 m)

deep. Above them soar lines of towering peaks. While thickets of shrubs cluster in the relative shelter of the valley bottom, the neighboring mountain heights are capped with ice and snow. Bare cliffs and scree predominate on the steep valley slopes, and in some places loose rock deposits dumped by ancient glaciers carpet the valley floors. Heavy falls of rain and snow are deposited on the western mountains, especially during

Thickly furred *against the cold, the snow leopard lives among remote crags and ravines, where it hunts mountain goats.*

The long-tailed *marmot lives beneath the soil in complex burrows and only emerges to feed.*

March and April, nourishing the icefields and rivers.

Some hardy animals make their homes on the rugged western slopes. Himalayan snowcocks forage for plant food on the edges of snowfields, making their nests among gray rocks and boulders against which they are well camouflaged. Siberian mountain goats are widespread and may forage at heights of 16,500 feet (5,000 m). Though sure-footed on even the steepest rocky slopes, they cannot always escape from rare and equally agile snow leopards, the elusive big cats of the central Asian mountains. The valleys of the western Pamirs are also home to brown bears, beech martens and mountain weasels.

Life is more difficult still in the eastern Pamirs. Though the terrain is gentler, with broad rounded surfaces and snowbound peaks rarely rising more than 5,000 feet (1,500 m) above the surroundings, the plateau is consistently high, bitterly cold and extremely dry. Average midwinter temperatures are as low as 0°F (–18°C) and yearly precipitation can be below 2½ inches (60 mm), giving the plateau the climate of a cold desert. Much of the ground is bare or lightly spotted with wormwood, cushion plants and steppe grass.

Animals are accordingly few, but admirable in their resistance to hardship. Among the species of mammals that do brave the eastern Pamirs are the argali, magnificent wild sheep that find forage even in winter in the broad valleys, and two well-furred rodents, the long-tailed marmot and the long-eared pika. Several of the birds that live here are those adapted to survive in similar harsh conditions farther to the east, among them Tibetan snowcocks and Tibetan sandgrouse.

Several high lakes enhance the unconventional grandeur of the eastern Pamirs and enliven its wildlife community. The largest of these,

Karakul in the northeast, is about 13,000 feet (4,000 m) above sea level. Though it freezes over in winter, this salt lake supports enough aquatic life in summer to attract nesting terns, gulls and waterfowl in search of food. Bar-headed geese make their nests on islets in the lake. They share their remote havens with colonies of brown-headed gulls, migrants from the coasts and lowlands of India, for which Karakul is the most westerly

Lake Karakul occupies a giant depression high in the eastern Pamirs. The glint of white on its margins shows that it is a salt lake.

breeding site. Strange it may be, but in a land of such contrasts as the eastern Pamirs, it seems fitting that a seagull should choose to breed far inland among some of the highest mountains in the world.

GLACIAL PEAKS AND TROUGHS

Glaciation has left its mark in the dramatic landscapes of the western Pamirs. Erosion by ice has eaten into the mountaintops, creating sharp-ridged, pointed peaks. Massive glaciers have gouged long trough-shaped valleys.

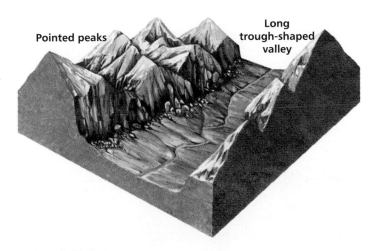

Pointed peaks

Long trough-shaped valley

ROYAL CHITWAN NATIONAL PARK

"Lush greenery, picturesque rivers and lakes, and the distant panorama of the Himalayan peaks"

Hot, humid and with an abundance of wildlife, Royal Chitwan National Park was once a hunting playground of nobility and royalty. Today, it is one of Asia's richest preserves and a vital refuge for

The mighty snow-capped Himalayas look down on the wet lowland wilderness of the Chitwan Park.

rhinoceroses, tigers and other critically endangered animals.

Chitwan is the best-preserved part of a belt of jungle and swamp that used to run the entire length of southern Nepal. Lying at the base of the great Himalayan chain, beyond its impressive foothills – the Mahabharats – these lowlands soak up both cascading surface water and seepage below ground from the rains that sweep the mountains. Water abounds, forming permanent rivers and marshes and inundating seasonal flood plains during the torrential monsoon peak. The air is stiflingly hot in the summer months, often reaching 100°F (38°C).

The park occupies a central section of this narrow belt. The broad, island-studded Narayani River

and its winding tributary, the Rapti, form its northern boundary, while the Indian border and another tributary, the Reu River, define the southern frontier. The broad flood plains that surround the rivers lie at about 500 feet (150 m) above sea level. They still have their natural landscapes – swaying rushes and tall grasses on the damp, level ground and strips of forest lining the banks of watercourses. Across the interior of the park rise the Siwalik Hills, a low range in places reaching 2,500 feet (750 m) in altitude.

With its lush greenery, picturesque rivers and lakes and the distant panorama of the Himalayan peaks, Royal Chitwan National Park has a beautiful, tranquil setting. Each of its varied landscapes has a special character of its own, and between them they provide habitats for a delightful diversity of animal life, including about 400 species of birds and at least 70 of butterflies.

The forested hills, deep and inaccessible in the heart of the park, are a rare preserve of seasonal

> *"Flood plain grasslands are one of the last strongholds of the Indian rhinoceros – a single-horned giant"*

tropical forest. This type of forest faces marked climatic changes through the year. Copiously watered during the monsoon, the trees then endure several months of dry season.

The forest habitat that develops differs from tropical rainforest in key respects. Many of the trees, for example, shed leaves during the drought, leaving their crowns bare, and at ground level the forest tends to have

a dry, grassy undergrowth. The interior forests of Chitwan are dominated by sal – magnificent, straight-trunked hardwood trees up to 100 feet (30 m) tall – although chir pine becomes common on the hilltops. The dense habitat the trees create provides sustenance and shelter for rare forest animals, among them sloth bears – curious, long-snouted bears which feed on ants and termites – and herds of gaur – heavily built wild cattle.

Many of the creatures that dwell in the extensive sal forests also occur in the riverside belts of forest on the flood plain. Here a mixture of acacia, shisham, kapok and other tall trees is characteristic. The towering kapok produces a marvelous display of fragrant red flowers in the spring.

Like the interior forests, these belts harbor numerous birds, the more conspicuous among them being Indian peafowls, hornbills, cuckoos and woodpeckers. Two kinds of monkeys, the rhesus macaque and the common langur, are numerous, as are wild boar and axis deer.

Big, boldly decorated and noisy, the great Indian hornbill hops from branch to branch, plucking fruit.

Rhinoceroses, their horns removed to deter poachers, wallow in the heat of the day to keep cool. In Chitwan they have benefited from the protection of guards and from the conservation of their favored marsh and grassland habitat.

Both boars and deer regularly move out of the riverine forest onto the grasslands. The flat savannas still provide excellent cover, however, because they are dominated by tall, dense grasses. Where elephant grasses predominate, the vegetation may be 20 feet (6 m) high, big enough to conceal not just the stately sambar deer but also the Indian rhinoceros, the second largest of the world's rhinoceros species. Chitwan's flood plain grasslands are among the last strongholds of this single-horned giant. About 350 rhinos, one-quarter of the total remaining population in the wild, live in Chitwan.

All grazing animals other than rhinos have to be wary of attack from tigers. There may be 100 Bengal tigers in the park, most concentrated on the flood plains, where they roam through forest, grassland and marsh

in search of game. The marshy beds of rushes and sedge that fringe the rivers, pools and oxbow lakes are also home to semi-aquatic predators such as the marsh crocodile and the Indian python.

For sheer numbers, birds are the most spectacular denizens of the marshes and lakes. Ducks, ibises, egrets, bitterns, herons, storks, hawks and kingfishers search the waters for their various foods such as plant matter, invertebrates, frogs and fish. Adjutant storks and black-necked storks are the largest species, ospreys and fishing eagles the most dramatic as they swoop to the surface to snatch fish in their talons.

Some of the birds present in the park during the winter are migrants, retreating from harsh conditions on their breeding ranges farther north in Asia. The striking Brahminy duck is

one of the most abundant, appearing in large numbers on the pools and open rivers.

The big rivers are also home to two rarer but more spectacular creatures. The gharial is a peculiar crocodile, up to 23 feet (7 m) long but with narrow, elongated jaws. The almost dainty snout is excellent for catching fish – minimizing water resistance when the reptile makes its sideways swipe – but is not powerful enough to cope with bigger prey. The Ganges dolphin, one of the few freshwater dolphins in existence, also has a long, toothy snout for grasping fish. It haunts the deeper stretches of water, principally in the Narayani River, and is occasionally spotted as it breaks the water surface. Like so many other rare creatures, dolphins and gharials find precious refuge in this untamed fragment of wild Asia.

THE SUNDARBANS

"Here, where land and sea merge, is a place so wild that tigers still rule its forbidding interior"

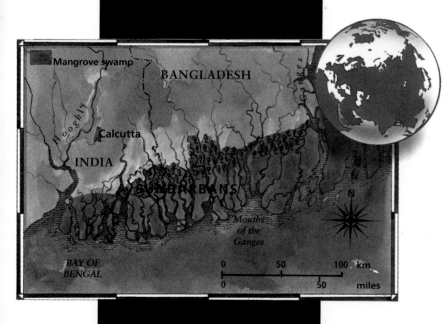

The dense, impenetrable jungle that lines the northern Bay of Bengal is no ordinary forest. At times, beneath the thick foliage there appears a tangle of starkly exposed roots in an ooze of soft mud, divided into islands by shallow, brackish creeks. At other times, the roots, mud and creeks are hidden beneath submerging sea water, leaving only trunks and leaves seemingly sprouting

Built of deposits of silt, clay and sand and shaped by the action of both rivers and tides, the low-lying landscape of the Sundarbans blurs the distinction between land and sea.

from the brine. This jungle swamp of the tidal zone is the Sundarbans, the biggest mangrove forest in the world. Here, where land and sea merge, is a place so wild that tigers still rule its forbidding interior.

The Sundarbans owes its existence to the great rivers – the Ganges and the Brahmaputra – which, along with other mingling waterways, have built up an enormous delta across southern Bangladesh and adjacent India.

Mangrove plants *help to create new land by trapping sediments among their exposed roots. The more silt that accumulates, the higher a patch of swamp becomes.*

Copious sediments from the Himalayas have been deposited as the rivers slowed on their approach to the sea and divided into a maze of channels. The seaward margin of the delta forms a tidal plain, a network of channels, mud flats and islands low enough to be inundated by the bay waters when the tide is high. This twice-daily incursion of salt water defines the character of the Sundarbans and sets it apart from the rest of the delta.

The greater adjutant stork, which stands up to 5 feet (1.5 m) tall, is a scavenger. Animal carcasses make up much of its food, but it also preys on small creatures such as frogs and fish.

"The bulk of the mangrove roots spread beneath the mud, but from them hundreds of rigid projections poke up above the ooze like miniature snorkels"

Stretching for 185 miles (300 km) between the estuaries of the Hooghly River in the west and the Meghna in the east, and reaching at least 18 miles (30 km) inland, the Sundarbans is not just the most extensive but also one of the most varied mangrove belts in the world. Though the drainage pattern of the delta was laid down by river flow, tidal water has taken over to different degrees, creating marked variations in salinity.

In temperate regions of the world, large tidal plains naturally turn into salt marsh. But in the warmth and strong sunshine of the tropics, the habitat that develops is much more lush. Instead of low-growing salt-tolerant plants, a mixture of evergreen shrubs and trees develop, known generally as mangroves. All are specially adapted to cope with the testing conditions of the tidal swamp – repeated immersion in salty water and a soil that is extremely low in oxygen. Special glands on the

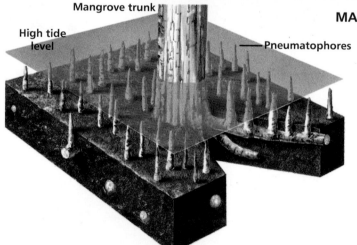

Mangrove trunk

High tide level

— **Pneumatophores**

MANGROVE PLANTS

At low tide, root projections, or pneumatophores, from the sundari mangrove tree are left poking out of the mud. They absorb air through their pores, and oxygen passes down to the roots beneath.

thick, waxy leaves secrete excess salt from the plants, and tiny openings in the bark of the exposed roots absorb oxygen from the air during low tide.

The roots of mangroves are their most extraordinary features. Some of the species in the Sundarbans have peculiar stilt roots that sprout from the trunk and curve down to the surrounding mud, encrusted with oysters and barnacles. In others,

including the widespread sundari that gives the swampland its name, the bulk of the roots spread beneath the mud, but from them hundreds of rigid projections poke up above the ooze like miniature snorkels.

Mangroves are great builders of land. Their roots trap sediments, and as the silt accumulates, the ground beneath the mangroves rises, lessening the degree of tidal immersion.

Throughout the Sundarbans there are patches of soil too high now to be flooded by the normal tide, especially inland where the swamp merges into monsoon forest.

Since different mangrove species have varying tolerances, both of levels of salinity and of flooding, the vegetation of the Sundarbans is far from uniform. Sundari, for example, is the most widespread of some 20 mangrove trees in the region. Since it prefers lower salinity, the plant grows tallest – up to 100 feet (30 m) – along the river estuaries in the east. Salty zones to the west tend to have shrubby mangroves, dominated often by the stilt-rooted goran. Low, deeply flooded areas are usually reserved for slender keora mangroves, while hantal palms may appear in dense groves on high, drier ground.

The mangrove world of the Sundarbans sustains and protects a teeming community of wildlife. At low tide, crabs are the most conspicuous creatures. Fiddler crabs sidle tentatively over the mud, picking at debris and darting back to their holes if danger threatens. Each male fiddler is encumbered with an oversized claw that it snaps loudly on occasion for self-advertisement. Mudskippers are equally curious denizens of the swamps. These small fish can breathe out of water and move over the mud using their front fins as limbs.

When the tide rises, the creeks overflow across the mud and roots. An array of creatures, tolerant of the brackish water, advance into the tangle to take their turn feeding on the rich organic matter of the swamp. Among them are shrimp, lobsters, fishes and river terrapins. Because of the abundance of both food and shelter, mangrove swamps also act as nurseries for a host of marine animals. Planktonic larvae fill the water, and of the 120 species of fish common in the Sundarbans, many

stay here only as juveniles. When full grown, they move out to open sea.

Though protected among the impeding roots and shallows from big predatory sea fish, mangrove fish still risk attack from above. Herons, egrets, kingfishers and fish eagles all hunt in the mangroves. When the mud is exposed, other kinds of birds, such as storks, ibises, sandpipers and stilts, move in to probe for the many

Monitors are large lizards, up to 6½ feet (2 m) long (below). They are good climbers and swimmers and hunt both in the trees and in water for small prey animals and birds' eggs.

The estuarine crocodile often basks on muddy banks, but is also perfectly at home in salt water (bottom). This fearsome giant often lurks with just its eyes and nostrils above the surface.

invertebrates hiding beneath. Flying insects fill the air above, including hordes of mosquitoes and biting flies. The insects provide some of the food for perching birds that forage and nest in the mangrove branches, among them shrikes, mynas and babblers. The trees are also shelter for the Sundarbans' monkeys – rhesus macaques. Some ground animals inhabit the drier parts of the forest but penetrate into the swamp at low tide. They include rodents up to the size of porcupines, groups of wild boar and axis deer.

Many of the ground predators also hunt for prey in the water. Among them are monitor lizards, pythons and fishing cats. The biggest predators of the swamps are those most feared but also most valued by poachers – the mighty estuarine crocodile and the Bengal tiger. For both, the tangled forests and muddy waters of the Sundarbans are a vital refuge.

FOOD GATHERERS IN THE SUNDARBANS

There are no permanent villages in the tangled swamps of the Sundarbans, but many thousands of people in India and Bangladesh rely on the region for their living. Harvesters travel into the swamps by boat and move around on foot at low tide as they collect honey and bees' wax and gather grass for matting, reeds for fencing and palm leaves for roofing material. Fishermen make camps in the area and work the waters for weeks on end.

But this wild place, with its healthy population of Bengal tigers, is a dangerous place to be. For years, the annual death toll from tigers has averaged an astonishing 50 to 60. In the past tigers were hunted ruthlessly, partly in response to man-eating.

Today the preferred approach is to reduce the dangers rather than the tiger population. Since the 1980s, some effective deterrents have been tried.

Human dummies have been placed in the forest and wired up to give the tigers a non-lethal shock. People working in the swamps have also been encouraged to wear face masks on the back of the head – tigers prefer to attack from behind unseen, and the masks seem to deter them. In 1987, when 2,500 people tried these masks, nobody wearing one was attacked.

Tiger numbers have been increasing in the Sundarbans in recent decades as a result of strict conservation, and there are now about 500 in total. This represents the largest surviving population in the world. Finding ways to protect people from attack by the animals without turning back the gains of conservation is crucial.

As they go about their work otherwise unprotected, honey-gatherers in the Sundarbans put their faith in a couple of simple ruses designed to ward off tiger attacks. Holding a stick over one shoulder appears to confuse stalking tigers as does wearing a face mask on the back of the head.

The Bengal tigers of the Sundarbans region have always been notorious as man-eaters. Though very few individuals selectively stalk and eat humans, some will attack on sight, and every year brings a list of fatalities. In 1988, 65 deaths were reported in a four-month period alone.

HIGHLANDS OF IRIAN JAYA

"A land of endless forest, green-clad hills and dramatic mountains, their summits gleaming with snow and ice"

Some of the wildest, least explored territory in all the tropics is contained in Irian Jaya, the western half of the great island of New Guinea. So impenetrable is its rugged, forested interior that, until the advent of air travel, tribes lived there without any contact at all with the outside world. Today, this easternmost province of Indonesia is still a land of endless forest, swamp and green-clad hills, with a chain of dramatic mountain ranges running east-west across its center. The highest peaks reach so far skyward that their summits gleam white with ice and snow.

The steep slopes of the highlands belong to a landscape that is young in geological terms. Indeed, the whole island of New Guinea is a relative newcomer on the world stage. Tectonic collision between two continental plates has literally pushed the island up out of the sea over the last few million years, buckling its central portion into a belt of mountains 100 miles (160 km) wide.

The deformed layers of rock that form most of the highlands today originated as sediments and limestone deposits from the ancient seabed. But these are interspersed with intrusions of volcanic rock, showing that the tensions in the Earth's crust also caused volcanic activity on the surface. There are no active volcanoes now in the mountains – although they do exist on islands close to New Guinea – but the upheavals that pushed the highlands up in the past are almost certainly still at work.

Rugged bare rocks, thin vegetation and glacier-hewn valleys in the central highlands of Irian Jaya contrast greatly with the familiar image of New Guinea as a lush, forest-filled island. The tallest of these mountains soar higher than Mont Blanc and are topped with permanent ice caps.

The whole region is frequently rocked by earthquakes.

The foothills, inner valleys and peaks of the Irian Jaya highlands experience widely differing climatic conditions. The contrasts are most sharply defined along the southern rim, where the land rises abruptly from lowland plain to the highest of the peaks. These include Jaya at an altitude of 16,500 feet (5,030 m), Daam at 16,140 feet (4,920 m), Mandala at 15,620 feet (4,760 m) and Trikora at 15,580 feet (4,750 m). From the hot, forest-blanketed lowlands, where temperatures average 80°F (27°C), conditions gradually cool with rising altitude at a rate of about 11°F (6°C) per 3,300 feet (1,000 m). Hence night freezing is common on the high ground, and snow regularly falls in the mountaintops.

The summits of Jaya, Daam and Mandala are all high enough to maintain small but permanent ice caps. These rank as some of the most dramatic of Irian Jaya's natural surprises, since the nearest icefields are in New Zealand and the

> "One of the few places in the world where new species of mammals are still being found"

Himalayas, more than 3,000 miles (5,000 km) away. Immediately below these ice caps and across the lesser summits, the landscape becomes one of bare rock and windswept tundra; below that, open, alpine-type meadows stretch across unexpectedly flat mountain plateaus. During the ice ages, most of these landscapes were glaciated – a past evident from the presence of lake-filled depressions, deposits of rock fragments known

as moraines, and other glacial landforms. The retreat of the ice may still be occurring, according to comparison with records from the last century. In fact, ice did not completely disappear from Mount Trikora until the 1940s.

The vegetation of the alpine zone varies considerably with the lay of the land. Tussock grasses and small herbs predominate in many places, with bogs developing on the flatter stretches. In more sheltered sites, and increasingly as altitude drops, there appear luxuriant tree ferns and shrubs, including a rich variety of rhododendrons. Rats, bandicoots and the curious long-nosed echidnas are among the few types of mammals that forage over these high, open habitats. Though the bird life is similarly sparse, several species have their sole home in these remote heights, among them snow mountain quails and orange-cheeked honeyeaters.

Below about 12,300 feet (3,750 m), the meadows and shrubland start to blend into forest on the steep mountain flank. At first, the forest is thin and composed of stunted, elfin trees, many of them conifers and myrtles. Gradually, with descending height, more favorable conditions cause the forest to become denser and richer in tree species. Still, however, there are many gaps in the canopy, often caused by minor landslides on the steep slopes. The spaces are quickly colonized by tangles of bamboo. Throughout this montane forest, humidity is high – often reaching 100 percent in the cool night air. The moist trunks and branches of trees are

The long-nosed echidna is covered with protective spines similar to those of a porcupine. It searches the forests and mountain grasslands mainly for earthworms, which it hooks onto bristles on the front of its tongue and pulls end-first into its narrow "beak."

The dwarf cassowary *(above) dwells in the mountain forests and feeds mainly on fruit. When threatened or cornered, it is likely to attack, striking out with vicious claws.*

A delicate film of spiders' webs stretches between rhododendron bushes below Mount Jaya. The webs are nurseries, giving protection and food to swarms of young spiders.

typically coated with moss, ferns and orchids, and it is common for the forest to be swathed in cloud during the day.

Cloud seldom forms on the warm slopes below 6,600 feet (2,000 m), and it is on these lowest parts of the ranges that the forest starts to resemble lowland tropical rainforest in its structure and diversity. The trees are tall, the canopy closed. "Monkey puzzle" araucarias are among the most distinctive of a wide variety of trees, and there are numerous species of palms and creepers.

From the tree line to the lower slopes, the forests that sweep down the mountainsides burgeon with remarkable creatures. Here giant stick insects and dazzling birdwings – the world's biggest butterflies live. The mountains are the stronghold

of the powerful New Guinea harpy eagle, which often flies below the tree canopy as it hunts for forest prey. Colorful lorikeets and parrots search for fruit in the treetops, their place taken at night by foraging fruit bats.

Much of the animal life of New Guinea has reached the island from nearby Australia rather than from Southeast Asia. Consequently, though the mountain forests have no native deer, monkeys or squirrels, they are home to various marsupial mammals, among them cuscuses, possums, forest wallabies and tree kangaroos. Since many of the tree-dwelling species are nocturnal, they are rarely seen and little is known of their lives.

Given the rugged nature of the terrain and the paucity of scientific investigation in the region, it is perhaps not surprising that these

forests are one of the few places in the world where new species of mammals are still being found. One of the most recent was a variety of tree kangaroo discovered in 1989.

The mountain forests of central Irian Jaya also harbor cassowaries, bowerbirds and birds of paradise, which have their fellow species in northeastern Australia. Birds of paradise, though they are famous for their extraordinary plumage, are much more easily heard than seen as they utter their distinctive calls from high in the canopy. At least 10 species occur at varying heights on the mountain flanks. Halfway up Mount Jaya, a loud crescendo of strange, sputtered noises might sound like radio static, but it is a King of Saxony bird of paradise calling from its perch in this wild, moss-laden forest.

Huts in a Yali village, set in steep, forested terrain, show the compact, rounded shape and pole-and-thatch construction typical of the homes of highland peoples in Irian Jaya. Roofs hanging low over the walls and false walls behind the door openings help to keep out drafts when the mountain air becomes cool at night.

PEOPLE OF THE HIGHLAND VALLEYS

Deep in the seemingly formidable mountain barrier that runs the length of central New Guinea lies a series of broad, well-watered, upland valleys. The extent of these valleys was not fully appreciated by the colonial administrations that ruled the island until well into the 20th century, when exploration became more advanced. Neither was the remarkable secret they held. Here, hidden from the outside world, lived hundreds of previously unknown tribes. Some were hunters and gatherers, many cleared small plots in the forest, grew crops for a few years and then left the plots to regenerate. But some – among them the Yali and the Dani people of Irian Jaya – had turned great sections of the valleys into cultivated landscapes, with neat fields and thatch-roofed villages. Evidence suggests that the well-managed farming systems may have been in existence in the valleys for more than 5,000 years.

Today, with the creation of numerous airstrips, most of the estimated 600,000 indigenous highlanders of Irian Jaya have some kind of contact with the exterior world. In some cases, change has overwhelmed their traditions, but many others still lead lives close to those their ancestors have lived for centuries. The huts in which they dwell, the types of implements they use, the few garments they wear, and their customs and rituals follow tradition. Some tribes actively shun outside contact, and a few still engage now and then in intertribal warfare.

The Dani people have cultivated large stretches of central Irian Jaya for thousands of years (left). Their sophisticated farming system centers on the production of sweet potatoes and on pig-rearing.

Dani men set about building a hut using products of the upland forest – lumber, saplings, vines and palm leaves (above left). The large bolster of grass acts as insulation for a sleeping platform.

Gourds of various shapes are used as storage vessels. The "net-back," a type of headgear extending over the shoulders and back, doubles as a garment and a large bag for carrying crops, piglets and even infants (above).

KRAKATAU

"Scene of the most immense explosions ever recorded on Earth, which changed the face of the entire archipelago"

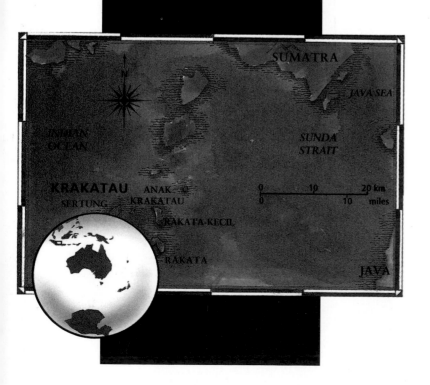

Midway across the Sunda Strait between Java and Sumatra lies an archipelago of four uninhabited islands. Set in a blue tropical sea, the outer three islands display the lush green hue of dense forest. Two are low in profile; the third rises to a sharp crest more than

Where Krakatau tore itself apart more than a century ago, a new volcanic island, Anak Krakatau has already risen from under the sea.

2,625 feet (800 m) high, its inner side forming a dramatically curving cliff. Facing the cliff is the fourth and smallest island – rugged and dark, with a conical mound rising near its center. A thin stream of smoke often trails from its summit.

Though the smoke barely impinges on the serenity of the islands, it is a hint that tranquillity does not always reign there. For these wild remote lands are the volcanic remnants of Krakatau, scene in 1883 of the most immense explosions ever recorded on Earth. The explosions were the climax of a cataclysmic eruption, one so violent that it changed the face of the entire archipelago.

Krakatau's turbulent history is one of repeated rebirth. In prehistoric times, there appears to have been a single giant volcanic cone on the site of Krakatau, built up out of the sea mainly from lava flows pouring from a subterranean reservoir of molten rock, or magma. Most of the top of the cone was destroyed in a huge

eruption – probably one referred to in a Javanese legend from the year A.D. 416 – leaving a great bowl-shaped crater, or caldera. Only parts of the caldera rim still projected above the sea, forming low islands.

Over the ensuing centuries, magma from the same underground chamber found its way up through one of these islands and developed a new volcanic cone, Rakata. Later, two more cones, known as Danan and Perbuatan, grew from the center of the caldera and merged with Rakata to form one island. Some 4½ miles (7 km) in length, with a distinctive profile, this was the main Krakatau island.

By the 1880s Rakata had ceased activity, and Danan and Perbuatan were mostly quiet. Much of the main island was under thick forest, as were its two low-lying neighbors from the original caldera rim, now known as Sertung and Rakata-kecil.

But in 1883, after several years of frequent earthquakes in the region, Krakatau came vigorously back to life. From late May to August of that year, cycles of eruptions from Krakatau sent enormous ash-laden clouds up to 7 miles (11 km) into the air. First Perbuatan, then vents on Danan, started to emit steam and ash with explosive force. On August 26 the activity came to a head. Increasingly loud explosions from the volcano could eventually be heard all over Java. Thick black ash clouds

doubled in height, and platforms of pumice from the volcano's solid outpourings floated across the strait.

On the morning of the 27th, the eruption reached its climax, with four exceedingly powerful explosions. The third, at 10.20 a.m., was so titanic that it dwarfed even these others. Its force has been estimated as 2,000 times that produced by the bomb

> "During the climax of the eruption two-thirds of the main island disappeared"

THE COLLAPSE OF KRAKATAU
When Krakatau erupted in 1883, Danan and Perbuatan may have caved in first. Rakata's northern slopes were then left without proper support, and this huge section of the cone sheared off and slumped en masse into the sea. In 1927 a new cone, Anak Krakatau, began to rise from the seabed.

dropped on Hiroshima in 1945. The sound could be heard more than 3,000 miles (4,800 km) away, and pressure waves from the blast were detected all over the world. Great boulders were hurled across the Sunda Strait. The ash cloud from the climax towered 50 miles (80 km) high and, as it spread, plunged the whole area into two days of darkness.

But the most devastating effect of the Krakatau eruption was a series of tsunamis, or tidal waves, which swept out across both shores of the strait and into the Java Sea, destroying entire villages and towns. At least 36,000 people lost their lives.

The biggest of the tsunamis overwhelmed the shores of Sumatra and Java shortly after noon on the 27th. It took another day for the eruption of Krakatau finally to run out of steam and, eventually, as the murk and floating pumice cleared, people were able to approach the volcano by boat and see what had happened. What they saw before them was astonishing.

The precise chronology of events is unclear, but during the climax of the eruption, two-thirds of the main island disappeared. Some of it was blown outward, along with huge amounts of pumice and ash from the underlying magma chamber, but most appears to have collapsed below the waves. So much magma had been ejected from beneath that the island had caved in,

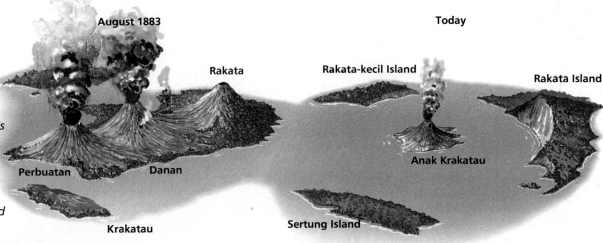

Thick, billowing clouds of steam and volcanic ash rise from the summit of Anak Krakatau (left). During Krakatau's biggest eruption, a total of nearly 5 cubic miles (21 cubic km) of matter was hurled into the air.

August 1883

Rakata

Perbuatan   Danan

Krakatau

Today

Rakata-kecil Island

Rakata Island

Anak Krakatau

Sertung Island

creating a new caldera within the old one, with its floor 1,000 feet (300 m) under the sea.

The collapse of the main island was the most dramatic – but not the only – reshaping of the Krakatau archipelago in 1883. Months of heavy ash falls had smothered Sertung, Rakata-kecil and the remaining southern face of Rakata, accumulating like sandbanks around their old shores, and enlarging Sertung in particular. Meanwhile, Krakatau was preparing for its next rebirth.

In 1927 new volcanic stirrings were detected from the caldera floor. Within a few years, a cinder cone had broken the surface of the sea, and it has continued smoldering, erupting ash and growing ever since. This is the small island in the center of the Krakatau archipelago today. Young, yet already topping 625 feet (190 m) above sea level, it is known as Anak Krakatau, or Child of Krakatau.

The island is for the most part barren. The frequent minor eruptions and the bare surfaces of cinder and ash are hardly welcoming to plants and animals. Yet even there, some plants are managing to grow. A few migratory birds stop to rest on the island, and seabirds feed around its shoreline. If the volcanic activity were to cease, it is certain that more variety of life would become established there. Rakata, Sertung and Rakata-kecil are proof of that.

All plant and animal life on the three outer islands was snuffed out and buried in 1883. Sterile ash covered all. Yet within a few years life, with its remarkable powers of recolonization, was returning to

Land-based animals such as this reticulated python have managed to recolonize the Krakatau archipelago since the great eruption. Their ancestors may have been swept ashore on rafts of floating debris.

On the harsh, *seemingly sterile surface of Anak Krakatau, a few hardy plants have taken root. Grasses, ferns and casuarina bushes all grow on the lower slopes.*

The summit of *Anak Krakatau is an unstable pile of cinder. Year after year, lava, rock and ash are pushed out of the vent of the volcano during eruptions, raising its cone ever higher.*

reclaim them. Hardy pioneering plants like grasses and casuarina appeared first. Many of these prepared the ground – changing its structure and building up soil as they grew and died – and the plants that followed could grow more easily.

The first trees began to spring up, providing more chances for insects, birds, bats and other creatures to find a suitable habitat. By the time Anak Krakatau was emerging, more than 270 species of plants had returned to the outer islands. At least 27 resident landbirds could be seen on Rakata, among them flycatchers, kingfishers and pigeons.

With the rich rainforests of Sumatra and Java only about 25 miles (40 km) away, it was inevitable that certain forms of life would soon reach the islands. Many plant seeds were carried there in the stomachs of birds. Others floated across the sea or drifted in the wind. Butterflies and other flying insects may have made landfall after being swept out to sea on strong winds. Some rodents and

reptiles appear to have floated across by accident on logs and rafts of floating vegetation. Today, the islands have a fairly rich fauna, including pythons, geckos, monitor lizards, rats and fruit bats. In both form and composition, the vegetation in which they now live closely resembles that of a typical rainforest.

But the evidence of the islands' violent past is never far away. Here and there, where the superficial soil layer has been disrupted, it is easy to see the powdery ash left by Krakatau's last explosion. Ultimately, life's hold on these wilderness islands might once again be extinguished. Perhaps Anak Krakatau will see to that.

THE KIMBERLEY PLATEAU

"Soft sandstone hills have been turned into natural sculptures"

S ummer comes suddenly to the Kimberley Plateau, in the far northwest of Australia. By mid-November, the land seems to have contracted after the long months of "the Dry." Red rock shimmers in temperatures which since July have increased to as much as 120°F (49°C), and billabongs – the

Tiger-striped with silica and lichen, the eroded domes of the Bungle Bungle Range glow magically as the sun rises over Kimberley.

pools of water left in empty river beds – provide a last sanctuary for freshwater crocodiles and sawfish. Boab trees living off the moisture in their swollen trunks have shed their leaves in the heat.

Offshore, tropical storms gather force, heralded by banks of bruised clouds and hot winds. They sweep inland over the fragmented shore and mangrove swamps, over stretches of mallee scrub and grassland, and up onto the plateau, where the first of the summer deluges begin.

Paradoxically, although the Kimberley Plateau consists of a whorl of arid mountain ranges flanked to the south and east by the Great Sandy and Gibson deserts, it accounts for more than three-quarters of the water runoff in Western Australia. The summer rains – which average 12 inches (30 cm) a month – pour off the impervious crust of the rock and rush through deep fissures formed by millions of years of rainfall. The engorged Mitchell River, which has

been known to widen from 330 feet (100 m) to 7 miles (11 km) after a few days of rain, rises by as much as 56 feet (17 m). Saltwater crocodiles and stingrays can be found right up to the base of the plateau cliffs, more than 60 miles (100 km) inland.

The Aboriginal name for the Kimberley is "Wandjina," after its creators the Water Spirits, and water is almost wholly responsible for its extraordinary physical geography. The mountains sit on what is known as the Ancient Plateau of Australia, a stable bed of rock which existed long before life on Earth. Some 350 million years, ago these ranges were part of an oceanic bank of limestone and coral far bigger than the Great Barrier Reef. As the sea receded and tectonic forces pushed the mountains upward to heights of more than 3,300 feet (1,000 m), the rain began to do its work, rounding off peaks and scoring the rock with numerous crevices and riverbeds.

Until the Gibb River road was built in 1986 to transport beef from isolated cattle stations, the northern part of the Kimberley Plateau was completely inaccessible. Even now the dirt road peters out into nothingness around Mount Elizabeth, and few venture into the interior when faced with the prospect of searing heat and floods. Travelers stick to boat journeys up the coast, to what is known as gorge territory and to natural phenomena such as the Bungle Bungle sandstone formations.

Some animals, meanwhile, have adapted adroitly to the violent extremes. The wallaroo, a type of kangaroo, reduces water loss by having concentrated urine and no sweat glands. It can survive for two weeks without drinking, keeping cool by licking its forearms and panting.

It is easy to think of the Kimberley as a compact, isolated piece of territory; easy to forget that its 30,000 inhabitants are scattered across an area the size of the Netherlands. The variety of scenery is astounding. The deep red gorges of Windjana and Geikie, studded with gray-green eucalyptus shrubs, and the 2,500-foot (750-m) subterranean cavern at Tunnel Creek – where the Rainbow Serpent is believed by Aborigines to have entered the Earth after the Creation – slice channels through the fossilized limestone of the ancient coral reefs.

To the west, on the low savannas, the only relief from the flatness comes from ungainly boab trees – relatives of the African baobab – thought to have floated over from Madagascar in seed form 75 million years ago.

Swollen trunks of boab trees dot the dry savanna near the coast. With their own supply of water in the trunk, the trees can outlive the worst of the summer heat and winter drought, sometimes splitting open in the process.

Tropical storms and powerful ocean tides have fragmented the red coastal rock of the Kimberley Plateau and fashioned it into exotic shapes (left).

Eerily translucent, a ghost bat hangs from the roof of a cave (right). One of the most predatory of all bats, it feeds on birds, lizards and even other bats. It attacks by landing on its victim, wrapping it in its strong wings and delivering a single killing bite to the back of the prey's neck. The ghost bat's sharp teeth have earned its other common name of false vampire bat.

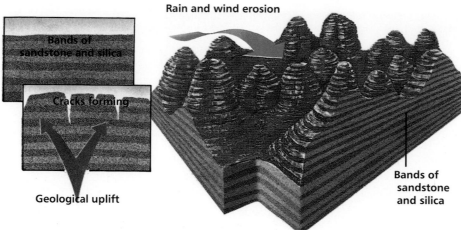

Rain and wind erosion

Bands of sandstone and silica

Cracks forming

Geological uplift

Bands of sandstone and silica

THE BUNGLE BUNGLE MOUNTAINS

Massive forces lifted layers of silica and sandstone on the Kimberley Plateau some one and a half billion years ago, leaving cracks and fissures in the vulnerable rock. Millions of years of wind and water erosion, chiefly from the flash-flooding caused by seasonal downpours, exposed these weaknesses and created the domes and ridges that make up the Bungle Bungle Range.

Miniature ranges of narrow termite mounds are built along strict axes to escape the worst of the sun. But even here water has taken a hand. Whiskey, Nugget, Blackfellow and Bottle Tree creeks, each a tiny thread on the map, are just some of the thousands of watercourses that vein the land, giving rise to small stands of eucalyptus and vine thickets wherever the water supply is permanent. These oases harbor exotic creatures like red-collared lorikeets, death adders and the purple-crowned fairy wren, one of Australia's rarest birds.

Finally, beyond the transitional coastal zone where mangroves survive in brackish, salty conditions – excreting salt from their leaves and supporting a wealth of anerobic life beneath their roots – there are the. wide mud flats and racing tides of the Indian Ocean and the Timor Sea. This is a treacherous coast – where water brings life inland, it brings death by the sea. Tides rising by as much as 36 feet (11 m) cover the mud flats faster than a person can walk, sweeping rays and ocean fish into the creeks where "salties" (voracious saltwater crocodiles) wait patiently. Boats caught unawares by whirlpools, vicious ocean currents or cyclones hurtling in toward land, have little hope of survival.

For this reason, the Kimberley coast is rarely seen by humans, except in photographs or the occasional film.

It is spectacularly beautiful. About a quarter of a million wading birds stalk the mud flats, leaving every year on a 1,250-mile (2,000-km) odyssey to Siberia for the breeding season. Mudskipper fish hop rides on fiddler crabs, and red mullet explode out of the water to escape from predators.

Southeast of the main plateau, outside the protective curve of the King Leopold and Durack mountain ranges and below the newly irrigated valleys of the Great Ord River, lies the most dramatic work of the Water

"Modern life has barely grazed the edges of the Kimberley Plateau"

Spirits – the Bungle Bungle Range. Wind and water have turned the soft sandstone hills into monumental natural sculptures, made more dramatic by the stripes of black lichen and orange silica inherent in the rock. On the cliff sides, and hidden in overhangs, are paintings done over thousands of years by Aborigines; tributes to the animals they hunted and the spirits who created them.

No one is allowed to climb the formations; one breach in the thin, hard skin of horizontal quartzite rock on the surface would expose the interior to erosion, and the Bungle Bungles would gradually disappear from the face of the Earth.

Modern life has barely grazed the edges of the Kimberley. There is a brief landscape of rolling pastureland dotted with sheep stations to the south of the Mitchell River, before the desert begins. In the east, millions of acres of grazing have been flooded in the Ord River Irrigation Scheme, which has succeeded in providing a consistent supply of water for agriculture. Aborigines continue to spear and net fish from their remote settlements on the coast. But the heart of the northwest, the plateau and its environs, is still empty; a haunting wilderness visited by the summer rains, baked by the sun and peopled with hardy animals and plants. Kimberley is still Wandjina: the land of the wind and the water.

The Windjana Gorge (above) was once part of a limestone and coral bank on the ocean floor.

In the rainy season, the dry plateau springs to life. Seeds sprout and for a time the flat dusty plain fills with the purple and green haze of flowering mulla mulla plants (right).

Brilliantly colored in ocher and gypsum, Wandjina figures cavort across a wall in the Dog Cave area of the Kimberley's Napier Range (above). Associated beings in the picture include a python, which represents the Rainbow Serpent.

Ancient paintings of the Wandjina – ancestral beings from sky and sea who formed the landscape and control fertility as well as the elements – still play a significant part in the ceremonial life of Kimberley Aborigines today (left).

ABORIGINAL ART
IN THE KIMBERLEY

The colorful, sedimentary rock walls of the central and northern Kimberley hold the secrets of the ceremonial life and rituals of the Wororra and Ngarinyin Aboriginal peoples. In tunnels, caves and overhangs, paintings in ocher and gypsum – some of them several yards long – have survived changes in climate and in sea levels for as long as 3,000 years.

There are two types of rock paintings in the Kimberley – the older, monochromatic figures known as "Bradshaw figures," after the European who first saw them, and the more famous "Wandjina figures," with stark white bodies, blank eyes and haloes of lines radiating from their heads.

The Wandjina came from the sky and sea, bringing rain and therefore fertility. They were often mouthless, and the lines around their heads may well signify lightning. They are frequently depicted with animal forms and powerful spirits such as the Rainbow Serpent or the Lightning Brothers. Repainted and overpainted by generations of artists, the Wandjina images are still relevant to Aboriginal culture today.

All Aboriginal rock paintings, however remote from modern settlements, are enmeshed in a system of tribal ownership. Repainting – which keeps their influence strong – may not be done by the owners, only by artist kin as dictated by tribal law.

Wandjina figures, such as these on a rocky overhang (right), retain their magic for centuries. They can be repainted to renew their spiritual strength or painted over with new designs, but only by specific artist relatives of their original owners.

CAPE YORK PENINSULA

"One of the most demanding and varied terrains on Earth"

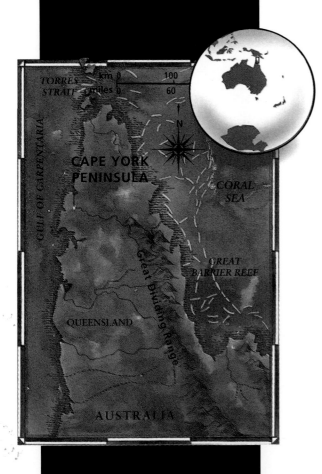

There is only one road to the northeastern peninsula the Australians call simply "The Tip." For six months of the year it is impassable: dry water courses fill with raging currents, rivers appear as if from nowhere and the dirt "washboard" track (so called because of its punishing, runneled surface) turns to mud under the

A strip of sand in a remote bay divides the forests of the Cape York Peninsula from the brilliant waters of the Coral Sea.

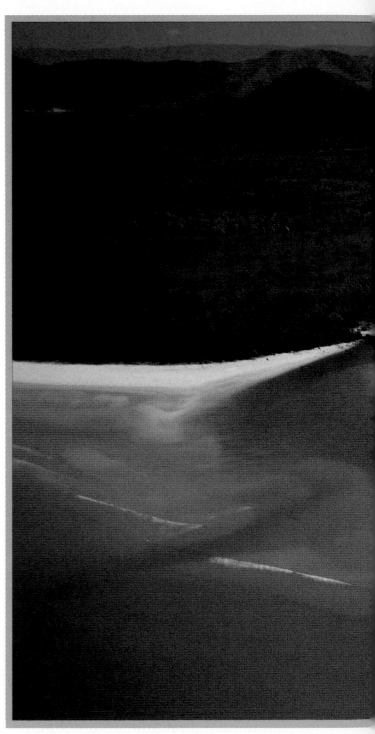

onslaught of tropical storms. Even during the drier months of June to September, it can take days to reach the isolated settlement of Bamaga, at Australia's northernmost point in the state of Queensland.

The Cape York Peninsula, with its 80,000 sq. miles (207,000 km²) of wilderness and its tiny population, occupies a special place in the urban Australian consciousness as the last frontier. The pockets of relict rainforest on the east coast are the sole remaining fragments of Australia as it once was – a land covered by millions of acres of dense forest supporting outsize, primitive plants and animals such as kangaroos. But anthropologically speaking, Cape York should be considered the first frontier, the point at which, around 40,000 years ago, the ancestors of today's Aborigines first crossed from Southeast Asia to populate a new and empty continent.

Today, few people head north over the Mitchell River to explore this remotest area of Queensland. Divided from Asia by the narrow waters of the Torres Strait and split along its length

by the fading spine of the Great Dividing Range, the peninsula is markedly different on either side of the mountains. To the west, the wide, tussocky savanna – grazing territory for the hardy cattle known as "scrubmasters" and home to creatures such as native bush cats and agile wallabies – ends in tangled mangrove swamps, desolate beaches and the shallow Gulf of Carpentaria.

By contrast, the narrow strip of land between the eastern foothills of the Great Dividing Range and the Coral Sea on the eastern coast of the peninsula is humid and heavy with vegetation, much of it a continuation of the great Indo-Malaysian rainforest. Giant buttressed trees – their central roots so shallow that the buttresses are vital to counter the weight of the canopy – are hung with lianas, orchids and clinging aroids (members of the lily family).

Every year between November and May, tropical fronts sweep across the peninsula en route to Asia, drenching the eastern forest, replenishing the southern wetlands, and bringing flash-floods to the western savanna. In January, temperatures can climb as high as 95°F (35°C).

The natural life of Cape York, as elsewhere in Australia, is partly the result of the continent's history. Australia was once part of the huge southern landmass known as Gondwana. When, about 160 million years ago, this supercontinent began to split into the separate landmasses of the southern hemisphere, the scattered descendants of Gondwanan species began to evolve in different ways. Among the fauna and flora which developed on the "Australian Ark" – finally isolated millions of years later – were flightless birds, marsupial mammals, the egg-laying mammals known as monotremes, and sclerophylls. These are trees that have developed thick, moisture-retaining cuticles on their narrow leaves. Best known of these is the eucalyptus, a common tree in the rainforests of the Cape York Peninsula.

One resident of the forest, the 6-foot (1.8-m) tall cassowary, a shy, flightless bird, is related to the African ostrich and the South American rhea – both fast-running, flightless birds. The group probably originated in Gondwana before it split up and only later evolved into separate species.

Others, such as birds of paradise, with their spectacular plumage, or the nocturnal sugar glider, a flying possum which skims from tree to tree by means of a parachute-like membrane between its fore and hind limbs, are Southeast Asian in origin. Native Queenslanders include Lumholtz's tree kangaroos – timorous

A sugar glider busies itself with its prey high in the forest canopy (left). This little creature can swoop from tree to tree, gliding on the extended membrane of skin between its legs.

Poised for flight, a giant tree frog clings to a vine, using suction pads on its fingers and toes (right).

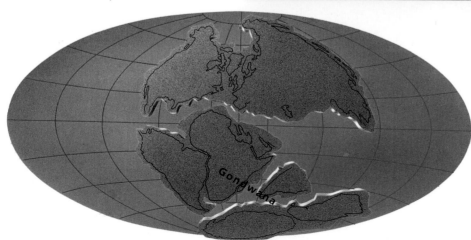

BREAK-UP OF A SUPERCONTINENT

The supercontinent of Gondwana began to fragment some 160 million years ago. As the continental masses slowly drifted apart, each still bore its original flora and fauna. Once parted, these developed into separate species. Marsupials, however, evolved after Africa and India had separated from the others but before the break-up was complete, so they are found only in South America and Australia and as fossils in Antarctica.

"Pockets of relict rainforest are the sole remaining fragments of Australia as it once was"

creatures that can drop 45 feet (14 m) from a tree perch down to the ground to escape from danger – and bearded forest dragons – a rare type of lizard.

The peninsula has never welcomed intruders. Its coast is sparsely punctuated with the ruins of gold and pearling towns – eerie testaments to human greed which are gradually being reclaimed by nature. The Aboriginal population has survived, exchanging its semi-nomadic family units for the settlements encouraged by federal government. Even so, the 2,500 Aborigines on the peninsula – including the Torres Strait Islanders – lead an isolated existence; most reserves and communities are far out of reach of the casual visitor.

In the sandstone escarpment at Split Rock, near the town of Laura, Aboriginal history is recorded in the natural "museum" known as the Quinkan Galleries. Dozens of rock art chambers, some 13,000 years old, lie open to the sky, the majority of them off-limits to outsiders. Dugongs, or sea cows – once a valuable source of food – swim across the walls with giant turtles and crocodiles, while scenes of hunting and gathering depict species and ways of life that came and went as sea levels changed.

Cape York's natural wealth has always been the biggest threat to its survival. The iron-rich laterite soil of the west already supplies the world with bauxite, and the ancient but comparatively unfolded peninsular rock contains colossal riches in the form of metals and minerals. The precious area of rainforest on the peninsula has shrunk by three-quarters since European contact. What remains is now under government protection.

But the very remoteness of this wild area may be its salvation. Cape York Peninsula looks likely to remain one of the most demanding and varied terrains on Earth – the first, and last, frontier of Australia.

SOUTH WEST NATIONAL PARK

"Low valleys are blanketed with the world's last great temperate rainforest"

Every few years, in the cool, remote forests of southwest Tasmania, there is a reported sighting of the Tasmanian tiger. It usually comes from a hiker on one of the long, strenuous trails in the 1,740-sq. mile (4,500-km²) wilderness known as the South West National Park, who believes he has glimpsed

Waves roll onto a lonely beach in Tasmania's South West National Park. Violent storms soak the land with 30 inches (750 mm) of rain a year.

the sloping rump and barred markings of one of the rarest animals on Earth.

If the tiger – in fact, a striped marsupial wolf long thought to be extinct – has survived anywhere, it would be here, among the mountains, rivers and watery sedgelands of Tasmania's loneliest region. The low valleys are blanketed with the world's last great temperate rainforest. Here, stands of gargantuan King Billy and huon pine tower over leatherwood, sassafras and blackwood trees, and clumps of ferns and velvet mosses cluster beneath them.

Temperate rainforest, unlike its tropical relative, is an airy place. Instead of the dozens of tree species with their oversized leaves and flowers and coiled lianas, one species of tree tends to dominate – in this case, the Arctic or tanglefoot beech. This is the only deciduous tree in the forest, which, according to tradition, waits for Anzac Day on April 25 before consuming the landscape in a blaze of autumn colors.

The Tasmanian devil was hunted to extinction on the Australian mainland by dingoes. This stocky marsupial is a scavenger rather than a predator. It has a big head and powerful jaws which can crunch through the bones of carcasses.

"A vicious type of scrub known as "horizontal" has tangled stems that grow vertically for a few yards, then shoot out sideways"

It may appear a gentle wilderness, but southwestern Tasmania is far from defenseless. Large areas have been saved from human intrusion by a vicious type of scrub known as "horizontal" – tangled stems that grow vertically for a few yards and then shoot out sideways, forming a dense tracery of branches which looks solid enough to walk on. Anyone foolish enough to try falls through into the undergrowth – often injuring themselves in the process.

These pockets of unknown territory, together with the more accessible parts of the forest, harbor two-thirds of the many mammal species unique to Tasmania, most of the 14 endemic birds, and several other animals and plants on the verge of extinction.

The island of Tasmania is large – 26,380 sq. miles (68,330 km²) – and shaped like an arrowhead, pointing straight down toward the South Pole from its position 100 miles (160 km) off the southeastern tip of Australia. The island was separated from the mainland 13,000 years ago, during the great melt from the last ice age, and the formation of the Bass Strait marooned dozens of species there to evolve in total isolation.

A stand of Tasmania's infamous "horizontal" plant blocks out the light and creates a false floor of branches to trap the unwary. The plant is almost impossible to cut through, even with an axe.

Today Tasmania shares little more than a continental shelf with its estranged parent. Unlike the dry, flat mainland, it is green and mountainous with a consistently wet climate. It lies directly in the path of the Roaring Forties, which bring rain and gales all year round, and during the hot summer it suffers torrential tropical storms from the northwest.

Perhaps because of its remoteness and atypical climate, Tasmanian wildlife tends toward the bizarre. The Tasmanian devil, a nocturnal forest-dweller with black and white markings and sharp claws, produces a blood-curdling scream when

cornered. There is a "living fossil"– a species of freshwater shrimp unchanged for 200 million years – still surviving in the frigid alpine tarns, and the improbable red velvet glowworm, which shoots a gluelike substance at its prey from protrusions on each side of its head.

The trees, though slightly less exotic than the animals, are equally impressive. Some of the huon pines – endemic to Tasmania – date from as early as 1,000 years before Christ and are only eclipsed in stature by the soaring trunks of moss-covered King Billies and the pale, lofty swamp gums. The latter, with a height of up

to 312 feet (95 m), is the world's tallest flowering plant.

Complex relationships have developed between fauna and flora. The Tasmanian honeyeater is a bird that feeds off the nectar of eucalyptus blossoms unique to the island. In doing so, the bird picks up pollen on its feathers, cross-fertilizing the trees and ensuring its supply of food for the future. Elusive orange-bellied parrots, of which there are now only about 200, congregate around a single type of tree called *Melaleuca*.

Though protected by law, the Tasmanian wilderness is desperately fragile. Even on designated trails through the open country, with its buttongrass swamps and delicate plants, the most careful walker can spread diseases such as root rot fungus, which attacks plants in moorland and dry eucalyptus forests. The alpine meadows contain more than 300 species of lichens, mosses and ferns which are easily destroyed, and a passionate campaign is under way to restore Lake Pedder, the deepest lake in all Australia, to its natural condition. The lake was flooded for hydroelectricity in 1972, destroying most of its rare fish.

But for the moment it seems that Tasmania's South West National

A duck-billed platypus dives in a freshwater stream, using its sensitive bill to help it find food such as shrimp. Seemingly defenseless, the platypus has a poison spur on each hind foot.

Park, with its eccentric animals, magnificent broadleaf forests and cold seas stocked with bluefin tuna and billfish, remains a safe haven for some of the stranger species on this earth. And perhaps in the deep recesses of the forest, protected by demanding terrain and fierce ramparts of "horizontal," the Tasmanian tiger still hunts its prey in peace.

POOR KNIGHTS ISLANDS

"Winds and strong seas have battered the islands' sides into jagged cliffs which rise straight out of the Pacific Ocean"

Few humans have set foot on Poor Knights Islands in the last century. Most of those who have are scientists camping out for a few days and nights on the small, craggy outcrops perched at the edge of New Zealand's continental shelf, staying only to

Sun streaming through underwater cave entrances at Aorangi Island illuminates a school of pink maomao fish.

observe the wildlife and the striking rock formations.

A combination of circumstances has made the Poor Knights a natural sanctuary for rare seabirds, insects and a number of reptiles. The islands usually appear on maps as two dots lying about 12 miles (20 km) east of the narrow spit of North Island known as Northland. In fact, there are four islands – Tawhiti Rahi and Aorangi (the two largest), Aorangaia and Archway – and some scattered rock stacks and pinnacles.

Their appearance is forbidding. Northeast winds and strong seas have battered their sides into jagged cliffs, which rise straight out of the Pacific Ocean to a height of 660 feet (200 m). The thin soil of the central plateaus, honeycombed with bird burrows, is surrounded by sparse bands of coastal forest. The climate is unexpectedly mild and subtropical. The islands are

ANCIENT REPTILES

The two largest Poor Knights Islands are famous for their populations of tuataras. These large primitive reptiles look like lizards, but are in fact the only surviving members of an ancient group known as beak-headed reptiles. Destroyed on the mainland by introduced predators such as cats and rats, tuataras thrive in the safety of the islands, feeding on insects such as wetas and beetles.

Tuataras are long-lived. They do not reach sexual maturity until they are about 20 and may live to be 100.

greatly sheltered by the bulk of Northland, and they lie in the path of a warm equatorial current which flows down from northern Queensland and around New Zealand's North Cape. It arrives at the Poor Knights Islands as the East Auckland Current, raising the water temperature by a few degrees and allowing a small pocket of exotic marine life to thrive in otherwise temperate waters.

Perhaps most important, islands such as Poor Knights are the only part of New Zealand that is rat-free. Their inhabitants are safe from the havoc wrought by these killers in the rest of the country. Despite the arrival of Captain Cook's ships, bringing pigs to trade with Maori villagers, no

"Lava tunnels and deep arches shelter rare black coral, and dazzling schools of fish are legendary among divers"

introduced predators reached the islands. The Maoris themselves, who hunted seabirds and their eggs for food, were massacred in a tribal raid from the mainland in 1830, leaving behind only the remains of stone walls and agricultural terracing that now provide homes for birds.

In spite of their small size, the Poor Knights Islands harbor several endemic species. Buller's shearwaters are large seabirds which breed in their thousands on Tawhiti Rahi, where the earth is soft and treacherous with their burrows. When the breeding season is over, the birds migrate as far as Alaska and Chile. They are an aggressive species, and other birds – like the rare Pycroft's petrel and the gray-faced petrel – have been forced to breed away from them in the relative security of Aorangi Island.

Bellbirds, too, breed on Aorangi where they eat the fruit of the hangehange trees, which grow perched on cliffs around the islands. They also eat insects from the tree canopy and nectar from the flowering xeronema lily, another local species.

Giant tree wetas and cave wetas – cricketlike insects which can cover 7 feet (2 m) at a bound – compete for living space in the sparse forests with a poisonous 24-inch (60-cm) long centipede. None of these could have survived if rats and other predators had reached the islands. New Zealand's largest gecko also lives on Poor Knights. A colorful 12-inch (30-cm) long lizard, Duvaucel's gecko, hides under stones or logs during the day and comes out at night to search for berries, nectar and insects to eat.

Tradition has it that when Captain Cook saw the islands in 1760, he christened them Poor Knights because they reminded him of a popular English dish of the time with that name – savory dumplings dipped in egg. But the Poor Knights started life more than nine million years ago as a huge and active volcano; today's tiny islands were once buried under 1,650 feet (500 m) of volcanic debris spewed down the western slopes of the crater. Gradually the cone wore down to mere remnants of rock, smoothed and shaped by the sea into their present form. The sweeping away of softer rock, leaving behind a skeleton of harder stone, created outlandish arches, caves and tunnels. Many are now far below the water level, providing a wide range of habitats and idyllic breeding grounds for marine life.

Where the cliffs emerge vertically from the ocean, they also plummet as much as 400 feet (125 m) to the floor of the continental shelf. Just beyond the shelf, the ocean plunges to far greater depths, producing strong swells and occasionally extremely dangerous currents.

Poor Knights Islands became a marine reserve in 1981. Landing on the islands themselves is now completely forbidden – the delicate balance of the ecology can all too easily be disturbed. Visitors are, however, allowed to explore the waters of the reserve. The "air-bubble caves" (submerged caverns with the roof higher at the back than the front, trapping a pocket of air), lava tunnels as much as 1,300 feet (400 m) long and deep arches are legendary among divers. They shelter clusters of rare black coral and dazzling schools of fish, such as maomao, boarfish and parrot fish. Great trails of red kelp – usually a feature of temperate, rather than subtropical, seas – grow to a size of 200 feet (60 m).

The warm equatorial current brings with it quantities of plankton – food for fish and seabirds alike. Mighty kingfish, weighing up to 80 pounds (36 kg), make their leisurely way through the rocks, caves teem with schools of blue and pink maomao fish and 40-foot (12-m) whale sharks swim ponderously above their shadows. Purple shore crabs, with powerful legs

and pincers, scale the rugged cliffs to forage for food among the trees.

Above ground, the only evidence of human occupation is the old 19th-century lighthouse on the north of Tawhiti Rahi and the remains of a Maori population of three or four hundred people. The Poor Knights Islands are left alone, cliff-bound and free of predators, to remain as they have always been, a tiny sliver of New Zealand's past.

Dawn comes to Aorangaia, one of the smaller Poor Knights Islands. Once buried under volcanic material, the remaining rock has been smoothed and weathered by the sea over time.

A mosaic moray eel and a spiny sea urchin strike a threatening note among the fairytale world of cup corals, seaweeds and colorful sponges on the ocean floor (below).

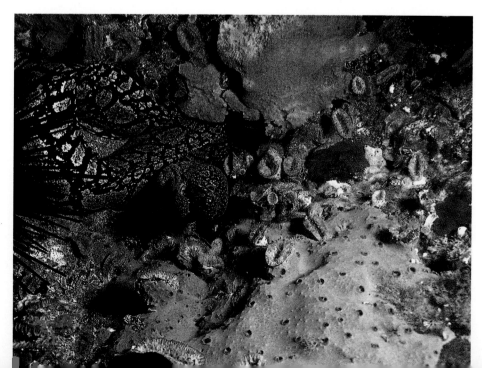

FIORDLAND

"Wild terrain with endless bush, few trails and rich animal life"

At the southwestern tip of New Zealand, the mighty alps and wild coast of South Island come to a dramatic end. Granite mountains plunge thousands of feet into the cold blue of the Tasman Sea, their slopes so precipitous that they are empty of vegetation. Fur seals raise their heads in sight of the shore, crested penguins dive for fish, and the probing fingers of the Marlborough Sound slide between sheer cliffs, penetrating deep into the haunting coastal region known as Fiordland.

In an already isolated country – New Zealand is about 1,000 miles (1,600 km) from its nearest neighbor – Fiordland is more remote still. Bounded on the north by a jumble of mountain ranges such as the Darrens, the Humboldts, the Stuarts and the Murchisons – their Maori names long since eclipsed by those of explorers and early Scottish settlers – it is wild terrain with endless bush, few trails and a rich variety of animal life. Due south of the Tasman Sea that borders Fiordland's coast, the Southern Ocean stretches away to the Antarctic ice cap.

Fiordland, which covers roughly 3 million acres (1.2 million hectares), was designated as a national park in 1952. The bulk of this beautiful area, named for its resemblance to the fiord coasts of Norway, is a battleground between land and water. After the last ice age, deep troughs gouged in the ancient, metamorphosed gneiss rock by glacial ice were flooded with sea water as the ice retreated. At the same

Mitre Peak, Fiordland's most famous mountain, towers 5,550 feet (1,692 m) above its reflection in the peaceful waters of Milford Sound. The Sounds – inlets carved into the rock by glaciers and flooded by the sea – combine with abundant rainfall to make this one of the wettest places on Earth.

At the heart of Fiordland, Lake Te Anau – once home to a mythical lost tribe of the Maoris – lies like a sheet of glass between the hills (above).

On a rare mist-free day, the icy waters of Sutherland Falls in Fiordland twist nearly 1,800 feet (550 m) down to the valley (right). These dramatic falls are the highest in New Zealand and among the highest in the world.

time, meltwaters poured eastward off the mountains and formed a chain of freshwater lakes dammed by glacial debris. These are now a natural barrier between the wild country and the rich pastures of the southern sheep farms. Surrounded and fragmented by water, the nation's biggest protected area seems to be slowly withdrawing from the solid fist of the mainland.

Even without its lakes and encroaching sea, Fiordland could claim to be one of the wettest places on the planet. Prevailing southwest winds dump 236 inches (6,000 mm) of rain on the mountainsides every year. Rain keeps the lowland temperate rainforest in a permanent state of humidity, replenishing lakes and feeding the hundreds of waterfalls that plummet from the cliffs.

New Zealand's isolation has had a profound effect on its flora and fauna. When it parted from other southern landmasses more than 70 million years ago, it took with it a number of flowering plants, but no animals. Living creatures have had to float, fly or swim to reach the islands, and the only endemic land mammals are two species of bats.

A richly forested land, with an equable climate and no serious land predators, New Zealand became a haven for birds. Many large, flightless species evolved – food was plentiful, and there were no enemies from which to escape. One of the largest was the now extinct moa which stood more than 12 feet (3.5 m) high. Today kiwis – New Zealand's national symbol – still forage for food in the Fiordland rainforest, using the nostrils at the tip of their long, curved bills to locate larvae and grubs. Diminutive, flightless birds with vestigial wings and a loose plumage of narrow feathers that resembles coarse hair, kiwis have survived in healthy numbers.

But for many birds, the arrival of human beings was catastrophic. First

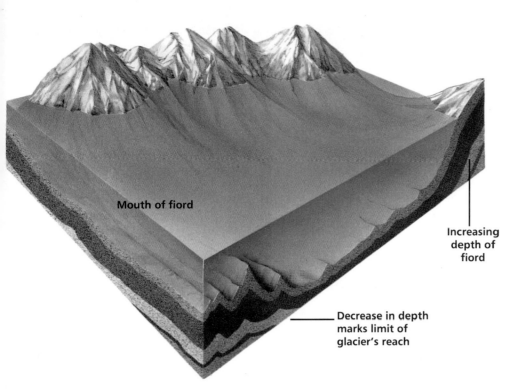

Mouth of fiord

Increasing depth of fiord

Decrease in depth marks limit of glacier's reach

THE NEW ZEALAND FIORDS

When the ice floes finally retreated from the southern tip of New Zealand some 10,000 years ago, they left behind a series of narrow, deep inlets that were eventually flooded by the rising sea. These inlets, or fiords, are typified by a steep-sided, deep channel where the ice scoured its way through the rock. As the fiord approaches the sea, its depth steadily increases. This is because the glacier that created the fiord grew in size as it was fed by side glaciers from tributaries. At the mouth of the fiord, there is a sudden rise in the sea floor, marking the limit of the glacier's reach.

came the Maoris, a thousand years ago, who trapped and speared moas, mutton birds and anything else they could catch for food. In 1770 Captain Cook sailed down the coast in the *Endeavour*, and in his wake came sealing and whaling ships with the most dangerous passengers of all – rats, dogs and pigs, which colonized the islands and found the clumsy native birds easy prey.

A century later, settlers hunted creatures they believed to be sheep-killers, including keas – olive green parrots with brilliant vermilion flashes beneath their wings. Keas can still be seen among the snow grass and mountain lilies of the high slopes. They are the only parrots in the world to live above the snowline and are almost certainly carrion-eaters. The birds that managed to survive the onslaught of the early European settlers retreated deep into the forest, along with the Maoris.

Today, no one hiking through the Fiordland bush will meet Maoris – the last sighting of the "lost tribe" of Ngati-Mamoe was in 1870 – or a predator larger than a feral pig.

> ## "Towering kauri pines, with their tremendous buttressed trunks and festoons of perching lilies, are hard to tell apart"

Certainly no one will die of thirst. The real danger is disorientation: once deep in the temperate rainforest, it is impossible to pinpoint any landmarks and easy to become hopelessly lost among the giant ferns and trees. Towering kauri pines, with their tremendous buttressed trunks and festoons of perching lilies, are hard to tell apart. Farther up in the subalpine forest of arctic beech and conifers, constant heavy rain and mist make the forest floor treacherous. Even the scrubland is thick with sword leaves, spear grass and the poisonous tutu plant.

As a result, large tracts of Fiordland are unexplored. In 1948, a naturalist named Dr. Geoffrey Orbell

was walking in the dense bush around the Murchison Mountains when he saw what he thought was a takahe – a flightless rail with a blue breast, green back and wings, and a large red beak, widely believed to be extinct. He had stumbled across a whole colony, which now numbers more than 300. The site was christened Takahe Valley and is one of the four areas in Fiordland National Park that may not be entered without a permit.

Another flightless rail found in Fiordland is the weka. These voracious birds hunt small creatures as well as feeding on grass and fruit and will even attack other ground-dwelling birds and steal their eggs. Wekas are still relatively common.

Much rarer are kakapos, the world's largest parrots, found only around the beech forests of the Cleddau Watershed on Milford Sound and on offshore islands. They are agile birds which, although almost flightless, hop and glide along established tracks to feed on grass, ferns and berries. Active at night, kakapos have sensitive bristles around the beak which help them find their way in the dark. Other birds on the danger list are laughing owls, of which there have been no confirmed sightings for 25 years, and kokakos, or wattle birds, which glide from tree to tree and produce a distinctive, sonorous call.

Fiordland's freshwater lakes contain some of the purest water in the world and some of the most awesome sights. On the western shore of Lake Te Anau, a labyrinthine cave system, well known to the Maoris, was discovered at around the time that Orbell found Takahe Valley. Deep inside is a cavern whose walls and roof are covered in glowworms –

"One of the wettest places on the planet where prevailing southwest winds dump a colossal amount of rain every year"

thousands of pinpricks of light scattered through the darkness.

Perhaps the most spectacular scenery in Fiordland is in Milford Sound, which winds inland for 10 miles (16 km), flanked on each side by the world's highest sea cliffs. At times the inlet is so narrow that it seems more like a rock tunnel than a fiord.

Conditions in the fiord are unusual. As the enormous volume of rain that falls yearly on the land filters down through the forests into the fiord, it passes through so much decaying vegetation that it turns the color of tea. When it arrives at the fiord, it does not mix with the salt water, being less dense, but forms a layer 10–13 feet (3–4 m) deep on the surface. The darker color of this top layer of water restricts the penetration of sunlight, and creatures that normally live in deeper water may be found in the first 130 feet (40 m) or so. Such species include sea pens, sea squirts and a huge colony of black coral.

On a calm, sunny day at the head of the sound, the bare walls of Mitre Peak and the spray of Bowen Falls swoop down to meet their mirror images in the still water. This is as far as visiting boats come. They take their photographs and head back to the sea, hoping for a glimpse of a dusky dolphin or even a sperm whale, and leave behind New Zealand's greatest untouched wilderness.

Giant tree ferns, known to the Maoris as mamaku, grow up to 50 feet (15 m) tall in the cool humid forests of Fiordland (left).

One of six species of parrots in New Zealand, the kea (below) uses its long upper bill to tear into carrion, fruit and insects.

The takahe (right) feeds on grass seeds and shoots. The birds are strictly protected, but are still seriously endangered.

GAZETTEER

*Besides the wild places
described in depth in this
book, there are a number of
other sites which can lay
claim to wilderness status.
A selection of such areas is
included in
the Gazetteer. Although most
of these wildernesses are
officially protected as
national parks and reserves,
many are still under threat.
All are priceless.
Their loss would be
an inestimable tragedy.*

POLAR REGIONS

1 Brooks Range ALASKA

Stretching in an unbroken barrier across the top of
Alaska, well inside the Arctic Circle, the peaks and
ridges of the Brooks Range reach from the Chukchi
Sea to the Canadian border and beyond. The
mountains represent the northernmost limit of
Alaska's trees. There are thickets of willow, spruce
and alder in the south-facing valleys, but the north-
facing slopes are bare. A profusion of streams and
rivers rush down from the mountains toward the
sea, crossing a coastal plain patterned with frost
polygons and meltwater ponds. The Beaufort Sea
once covered this plain, and fossilized coral heads
can still be found in the riverbeds.

The Arctic National Wildlife Refuge, which
includes the Brooks Range and the coastal plain,
covers almost 30,000 sq. miles (78,000 km²). Over
12,400 sq. miles (32,000 km²) of this has been
designated a wilderness. Wildlife in the park
includes black bears, polar bears and grizzlies, as
well as caribou, musk oxen, wolves and porcupines.

2 Northern Yukon National Park CANADA

This 3,860 sq. mile (10,000 km²) park, stretching
eastward from the Alaskan border, was established
as part-settlement of a land claim with the Inuvialuit
people of the region, who have an exclusive right to
hunt game within its borders and are involved in the
park's planning and management. Bounded on the
north by the Beaufort Sea, the park contains coastal,
mountain, tundra and taiga zones.

The 150,000-strong caribou herd that lives in
the park spends the winter in the southern taiga
foraging for food among the trees, sheltered from
the worst of the gales and blizzards. They move
northward as the weather improves in summer,
grazing on the tundra as the frosts melt, and
eventually crossing the mountains to calving
grounds in the foothills and plains of the coast.

3 Baffin Island CANADA

Some 1,000 miles (1,600 km) long and 500 miles
(800 km) wide, Baffin Island faces Greenland across
300 miles (500 km) of Baffin Bay and lies raggedly
across the entrance to Hudson Bay. This huge and
mountainous island is at the high eastern end of the
Canadian Shield, and where it drops abruptly to the
sea, it forms the deep fiords on the island's eastern
coast which frustrated early attempts to find the

North-West Passage. The mountains of eastern Baffin
reach altitudes of 7,900 feet (2,400 m) and more.

On the Cumberland Peninsula of the east coast,
Auyuittuq National Park is a 8,100 sq. mile
(21,000 km²) protected Arctic wilderness of mountain
peaks and valleys, glaciers, fiords and rocky shores.
Baffin Island is ice-locked for much of the year, but
has a rich and varied marine fauna, including ringed,
bearded and harp seals. An extensive migratory bird
sanctuary on the Great Plain of the Koukdjuak in the
west hosts the world's largest goose colony.

4 Greenland

The world's largest island, Greenland is covered with
a massive ice cap over 80 percent of its area. With
a relatively tiny human population, Greenland is
essentially one vast wilderness, and one-third of the
island, including all the northeastern coastlands, has
been designated a national park.

Most of Greenland's sparse wildlife, like its
human inhabitants, stays close to the coastal strips,
where there is at least a chance of seeing a little
greenery, in the form of mosses, lichens and grasses,
in the summer. There are herds of musk oxen and
reindeer in the national park area, as well as
wandering polar bears, with lemmings and Arctic
foxes at the other end of the scale. The fiords and
islands of the northeastern coast are home to
Greenland seals, common seals and walruses. On the
tundra of the Jameson Land peninsula on the east
coast, many birds, including long-tailed skua, red-
throated divers and whimbrels, breed along the rivers
and in the salt marshes. Large numbers of barnacle
and pink-footed geese also come here to molt.

5 Svalbard Islands NORWAY

Situated near the 80th parallel, 580 miles (930 km)
north of Tromsø, the Svalbard archipelago is high in
the Arctic. These nine main islands belong to Norway.
While commercial mining exists in part of the group,
45 percent of the total land area is protected by a
series of nature reserves and bird sanctuaries.

The name Svalbard means "cold coast," and
well over half of the area of these mountainous
islands is covered with glaciers and snowfields.
However, a warm current of the North Atlantic Drift
keeps open a channel through the ice to the western
coasts for most of the year. Edgeøya, third largest of
the islands, is a reserve favored by denning polar
bears. The islands are also a major breeding area for
seabirds such as guillemots, fulmars and kittiwakes.

6 South Georgia SOUTHERN OCEAN

The climatic influence of the southern ice cap
extends relatively far, compared to that of its
northern equivalent, and although the island
of South Georgia lies on the 54th parallel, it has
permanent snow and ice covering well over half of
its area. It is very mountainous, with 13 peaks over
6,600 feet (2,000 m). Snow falls for over 180 days
of the year. Gale-force winds are the norm, and in
winter the fiords are filled with broken sea ice.

Despite these rigors, South Georgia is something
of an oasis in the Southern Ocean, with slopes of
green mosses, lichens and tussock grass close to the
sea. The island is an important breeding ground for
a large number of Antarctic creatures, including huge
colonies of Antarctic fur seals and elephant seals.
Many species of birds come to South Georgia to find
ice-free nesting sites, including macaroni penguins
and several types of albatross.

7 Bird Island SOUTHERN OCEAN

Captain Cook named Bird Island when he discovered
it in 1775, close to the western end of South Georgia.

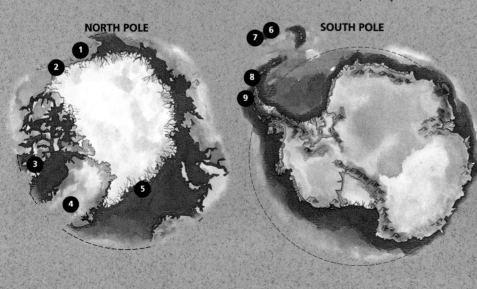

NORTH POLE

SOUTH POLE

Now, as then, the island's 370 acres (150 ha) can be almost entirely covered, shore to shore, with nesting birds of several species. A small British Antarctic Survey base houses eight or nine scientists, six of whom leave for the winter.

The albatrosses that come to Bird Island, ungainly on land, need grassy or mossy nesting grounds close to steep cliffs, where they can launch themselves easily into the air. The hordes of macaroni penguins prefer to lay their eggs on steep, bare rocky slopes. One particular rock face on Bird Island hosts 80,000 pairs. The only access from the sea is via a narrow channel, full of surging water and haunted by fur seals in search of a penguin meal. The island's petrels live in colonies of burrows. Non-burrowing giant petrels are the island's main scavengers, feeding on the carcasses of penguins that have fallen victim to seals, but also killing smaller birds and chicks.

8 South Shetland Islands SOUTHERN OCEAN

The Antarctic maritime islands of South Shetland are surrounded by pack ice in the winter. This chain of 11 main and many smaller islands is 336 miles (540 km) long. Most are volcanic in origin, and Deception Island, with its classic horseshoe caldera shape, is active, with hot springs and fumaroles. Tourists bathe in the steaming waters, but the bay has been known literally to boil, and major eruptions have occurred this century. King George Island and Livingstone are the largest in the group and house a number of stations manned by nations claiming Antarctic interests.

Close to the Antarctic Peninsula, the South Shetlands have extensive ice fields and glaciers, but also contain some large ice-free areas. Spectacular lichens flourish throughout the group, with some sea cliffs covered with a bright orange mantle. Large colonies of chinstrap, Adélie and gentoo penguins breed here, as do Weddell and elephant seals.

9 Brabant Island ANTARCTICA

Wild and inhospitable Brabant Island, situated close to the Antarctic Peninsula, has ice-covered mountain ranges rising to 8,200 feet (2,500 m) and just a few snow-free areas at midsummer. The island was not explored until the 1980s and was hardly visited before that. Many of the island's rocks show clear volcanic origins, with soaring 660-foot (200-m) cliffs of basaltic columns in places, patchworked with yellow and rust lichens. Elsewhere, in windswept and wet areas, bushy lime-green lichens color the rocks.

The only seals visiting the island in any numbers are Antarctic fur seals. Up to 1,000, mainly bachelor males, are there between January and March. Small numbers of Weddell, crab-eater and leopard seals also appear from time to time. The most common birds are chinstrap penguins, Antarctic fulmars, Cape pigeons, snow petrels, Wilson's storm petrels, blue-eyed shags, kelp gulls and Antarctic terns.

NORTH AND CENTRAL AMERICA

10 Round Island ALASKA, USA

A small extinct volcano of rocky basalt thrust out of the waters of the Bering Sea, Round Island lies 20 miles (32 km) west of Alaska's Nushagak Peninsula. There are no trees on Round Island, but despite its bleak situation, it is green with grass in the summer months and brilliant with flowers such as poppies, wild geraniums and forget-me-nots.

Round Island hosts some species of mice, and also red foxes which prey on them, but by far the most impressive inhabitants are the walruses, which cram the beaches between June and October. The 10,000 and more walruses spending the summer months on Round Island are all males, a vast bachelor herd that spends the time sunbathing and occasionally swimming off in small groups for a few days to feast on clams and crabs.

11 South Moresby National Park CANADA

Canada's Queen Charlotte Islands are the peaks of an underwater mountain range running parallel to the mainland coast of British Columbia. This archipelago of two large and 150 small islands, warmed by Pacific Ocean currents, receives massive annual rainfalls, and mists often cloak the evergreen rainforests that cover its mountain slopes. The South Moresby National Park, covering 570 sq. miles (1,470 km²) of the second largest island, was founded in 1987 after a long campaign conducted by the native Haida people against commercial logging.

In the dark forests of the park, there are trees possibly 1,000 years old. Some Sitka spruces tower up to 165 feet (50 m) in height, flourishing in the deep, black soil and humid conditions. Other forest giants include western red cedars, western hemlocks and Douglas firs. Wildlife species are often distinct island variants of mainland creatures, including the blue grouse, the northern saw-whet owl and, particularly, large black bears.

12 L'Eau Claire Wilderness CANADA

On the eastern shores of Hudson Bay in northern Quebec, L'Eau Claire Wilderness is a remote and inaccessible swath of rivers, lakes, channels, rapids and streams separating Lac à L'Eau Claire in the east from Richmond Gulf and Hudson Bay in the west. It is a transition region between the boreal forests of the south and the tundra of the northern Canadian Shield. The region has remained undeveloped due to its harsh winter climate and waterlogged terrain, but is hunted and fished by members of the Cree nation.

The wetlands are breeding grounds for mosquitoes, black flies and horseflies which descend in thick droves on any living creature. The depredations of the blood-sucking insects are capable of killing off old and weak caribou from the herds that migrate across the region.The waters and shores of l'Eau Claire are a paradise for many species of waterbirds, including divers, and favorite fishing and hunting grounds for ospreys and golden eagles.

13 Isle Royale National Park, Lake Superior USA

The largest island in Lake Superior, Isle Royale is 15 miles (24 km) from the Canadian shore, but is technically a part of Michigan, which is 50 miles (80 km) distant. It contains a unique ecological balance between moose and timber wolves, both of which immigrated from Canada, the moose having swum across around 1900, and the wolves having crossed the ice in the big freeze of 1949.

Isle Royale, 45 miles (72 km) long and between 2½ and 5 miles (4 and 8 km) wide, contains 27 lakes and provides a swampy lowland terrain where the moose can feed on water plants in the summer in an environment difficult for the wolves to penetrate. Making do with beavers and the odd moose calf in summer, the wolves' turn comes in winter, when the marshes and lakes freeze over. The two populations thus keep in balance, with about 1,000 moose and some 50 wolves, in three or four packs.

NORTH AND CENTRAL AMERICA

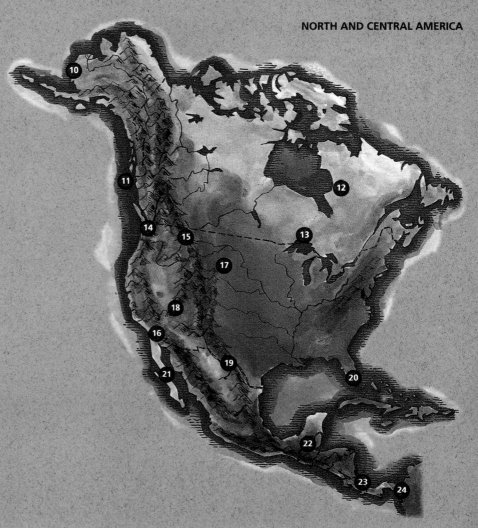

14 Olympic National Park
WASHINGTON, USA

This national park consists of a narrow coastal strip and a large central expanse of the Olympic peninsula, which occupies the most northwesterly point of the contiguous United States. The temperate rainforests of the park suck up the enormous annual rain and snow precipitations falling to the west of the Olympic Mountains. The forests are towering cathedrals of timber, ferns, and trailing festoons and carpets of moss and lichen. Some of the Douglas firs in the lowland forest attain heights of 250 feet (76 m) and more. Occasionally, brilliant flowers relieve the gloom of the forest floor.

Established in 1938 as a refuge for the Roosevelt elk, the park is a remnant of the rainforests that once cloaked almost the entire Pacific coast. In addition to Douglas firs, there are giant Sitka spruces and western hemlocks. A number of icy rivers rush down to the park's coastal strip from the mountains, some carrying spawning salmon in season.

15 The Great Burn IDAHO/MONTANA, USA
Straddling the Idaho-Montana border between the Clearwater and Lolo National Forests, the Great Burn wilderness is a mountainous forest region uniquely and indelibly marked by a great fire that occurred almost a century ago. In August 1910, a drought year in the Idaho and Montana Rockies, gale-force winds fanned a number of small fires and created an inferno which incinerated over 4,600 sq. miles (12,000 km²) of trees. Further fires in succeeding years destroyed seedling growth and prevented regeneration, creating instead large, permanent meadows where once there had been trees.

Now the area known as the Great Burn, covering about 390 sq. miles (1,000 km²), is a diverse mix of mountains, meadows and forested valleys, inhabited by elk, moose, deer and black bears, all of which thrive on the grasses and shrubs which grew up in the new meadows.

16 Yosemite National Park CALIFORNIA, USA
The stunning scenery of the Yosemite National Park in California's Sierra Nevada is in danger of attracting too many visitors, and numbers are being strictly controlled in order to preserve this vast natural wilderness of mountain cliffs, gigantic waterfalls, lakes and forests. Unknown to non-Native Americans until 1833, Yosemite became a state park in 1864 and a national park in 1890.

The park's dramatic rock walls, domes, peaks and spires owe their existence to a combination of movements in the Earth's crust and glacial action. A popular attraction is Yosemite Falls, consisting of an Upper Falls, the world's longest unbroken waterfall at 1,427 feet (435 m), and the Lower Falls beneath, continuing the drop for a further 330 feet (100 m). Yosemite's wildlife includes pumas, black bears, coyotes, mule deer and gray foxes. Pikas and yellow-bellied marmots, both rodents, live among the rocks. Most famous of Yosemite's trees are the majestic redwoods and giant sequoias, which can attain heights of 330 feet (100 m).

17 Badlands SOUTH DAKOTA, USA
The arid and eroded peaks, ridges and gullies of South Dakota's Badlands form a multicolored science fiction landscape that seems hostile and lifeless at first sight, but which is home to a wide variety of creatures and plants. Originating in layers of marine sediment, waterborne mud and volcanic ash, the Badlands today are the result of millions of years of erosion by rain and wind, a process which continues to shape the strange, sculpted terrain.

In and around the most arid gullies and ridges of the Badlands, the most efficient survivors are snakes, bats and rodents. Where the soil is more stable, plants such as prairie golden pea and buffalo grass help to bind it. Agriculture was banned in 1978; since then, prairie grasses have become more established, attracting numbers of the almost extinct pronghorn antelope, America's fastest mammal.

18 Grand Canyon ARIZONA, USA
Possibly the most famous of all American natural landmarks, the Grand Canyon repeatedly reduces new generations of visitors to awed silence. Created over two million years by the abrasive force of the Colorado River working against a giant upheaval in the Earth's crust, the Grand Canyon is today 276 miles (444 km) long. At its most extreme it is 1 mile (1.6 km) deep and 18 miles (29 km) wide. From Lake Powell at its eastern end to Lake Mead in the west, the canyon covers 2,000 sq. miles (5,200 km²).

Despite vast numbers of visitors who descend the steep trails and even raft down the Colorado's rapids, the Grand Canyon remains a true wilderness. The scale and beauty of its cliffs, ravines and empty spaces dwarf all human intrusion. The canyon's great depth creates a wide range of climatic zones, reflected in wildlife populations – from scorpions and lizards on the canyon floor, to gray foxes, herons and deer in fertile side canyons, and the unique Kaibab squirrel in the woods of the North Rim.

19 Big Bend National Park TEXAS, USA
One of the country's least visited national parks, Big Bend occupies over 1,250 sq. miles (3,240 km²) of wild country, embraced within a great curve of the Rio Grande which forms the border between Mexico and the United States. Terrains within the park's confines vary from mountain range and upland desert to lowlands and river environments.

The river banks are particularly rich in wild creatures, with mountain lions, rare Mexican wolves, bobcats, coyotes, collared peccaries and mule deer all coming down to drink. In the Chihuahuan Desert sector of the park live creatures adapted to arid conditions, such as scorpions, lizards, kangaroo rats and black-tailed jackrabbits. The Chisos Mountains soar in the southwest of Big Bend, with several peaks over 7,220 feet (2,200 m) inhabited by high altitude creatures such as Sierra del Carmen whitetail deer and kit foxes.

20 Everglades FLORIDA, USA
The subtropical wilderness of the Florida Everglades covers 33,000 km² (12,750 sq. miles) of the Florida peninsula and contains a mixture of habitats that includes pine forests, mangroves, marshy grasslands and shallow, slow-flowing rivers. The Everglades National Park covers 2,020 sq. miles (5,232 km²) of the tip of the Everglades. This strange, flat area slopes gently south and an 50-mile (80-km) wide river, 6 inches (15 cm) deep flows slowly toward Florida Bay in the rainy season, through expanses of saw grass.

Although threatened by developers and by the water demands of agriculture and the cities, the Everglades is still a refuge for several endangered species such as the Florida panther, the manatee and the loggerhead and green turtles. In the dry season, the alligators of the Everglades dig holes and dredge the clogged channels with their claws and snouts.

21 Baja California MEXICO
The great desert peninsula of Baja California runs parallel to the Mexican northwestern coast for some 800 miles (1,300 km), ranging in width from 30 miles (50 km) to 145 miles (230 km) and forming the western coast of the Gulf of California. The waters on each side of the peninsula are rich in fish and marine life, but the Baja itself is primarily arid, with a wildlife specially adapted to dry, desert conditions. The barren and waterless nature of the Baja has preserved it to a considerable extent from human settlement and exploitation.

Mammals on the Baja include the kangaroo rat, which does not need to drink water, as well as the jackrabbit, the wild burro and several bat species, including the fish-snatching bulldog bat. Mountain ranges run the length of the peninsula, and in the northernmost hills are pumas and wildcats. The birds of the Baja are concentrated around the coast, though the carrion-seeking buzzard thrives in the dry interior, as do several species of rattlesnakes and scorpions.

22 Lacandón Wilderness CHIAPAS, MEXICO
The Lacandón Forest is one of the last great stretches of primeval jungle in Mexico, bounded in the northeast by the Usumacinta River, which forms a border barrier between Mexico and Guatemala. Covering an area of some 2,200 sq. miles (5,700 km²), the forest contains several lakes and a number of tributaries of the Lacantum River, which eventually joins the Usumacinta to form the forest's southeastern boundary. Dugouts are the only feasible form of transportation within the forest, despite the crocodiles which can reach 25 feet (7 m) in length.

The Lacandón is a true rainforest. The tall trees are draped with orchids and bromeliads, and in the canopy above are noisy macaws and howler monkeys as well as toucans. At a lower level are shrubs and immature trees, and the forest floor itself is moist with rotting leaves and wood.

23 Amistead Biosphere Reserve COSTA RICA
Costa Rica has a particularly diverse flora and fauna which includes elements from both American continents. The country's Amistead Biosphere Reserve in the foothills and uplands of the Cordillera de Talamanca is in fact an assemblage of reserves and protectorates with a total area of 2,257 sq. miles (5,845 km²). The terrain includes glacial lakes, lowland tropical rainforest, subalpine paramo forest, oak stands, high altitude bogs and cloud forest.

The 215 mammal species within the reserve's varied habitats include tapirs, squirrel monkeys, jaguarundi, ocelots and jaguars. Endangered birds under protection include quetzals, orange-breasted falcons and harpy eagles.

24 Kuna Yala Reserve PANAMA
In direct contrast to most threatened indigenous peoples, the 30,000 Kuna Indians of Panama own the land on which they live and grow food. The reserve includes the 350 forested coral islands on which they live, and the narrow coastal plain where they use traditional agricultural methods, as well as hunting and gathering within the forested zone. The Kuna established their reserve in 1938, deliberately distancing themselves from the mainstream of Panamanian society.

In the 1970s, the Kuna were threatened by a new road used by slash-and-burn peasant farmers operating close to the reserve's boundaries. With the aid and advice of concerned organizations, they established the Kuna Wildlands Project, containing four categories of protected territory: a major area restricted to scientific research and controlled tourism; a Kuna agricultural reserve; a Kuna cultural region including the islands and coastal plots; and a restoration strip just outside the reserve, which acts as a buffer zone. The success of this project has encouraged other indigenous groups to follow suit.

SOUTH AMERICA

25 Ciénaga Grande de Santa Marta COLOMBIA

This mixture of wetland habitats on Colombia's Caribbean coast is situated at the mouth of the Rio Magdalena. Salt lagoons and mangrove swamps line the shore, and further inland is a complex of freshwater lakes, swamp forests and marshes, which can be completely inundated by the Magdalena when the river is in flood. The coastal region itself has a very low rainfall, and the floods are the result of snow melting in the mountains of the Sierra Nevada de Santa Marta.

This 195 sq. miles (500 km²) of wetlands and coastal lagoons is an important feeding, resting and breeding site for migrating waterfowl and resident species. These birds, the shellfish and crustaceans harvested by both birds and humans, and the mangroves lining the lagoons depend on delicate balances between fresh and salt water. Upstream irrigation canals have led to a reduced flow of fresh water from the Rio Magdalena, causing excess salinity in the lagoons and damage to the mangroves. Further damage to the whole ecology of the area will occur unless the freshwater flow is restored.

26 The Llanos VENEZUELA

The Venezuelan Llanos, part of the flood plain of the Orinico and Apure rivers, are lush grassy wetlands for half of the year and a dusty plain for the other half. Extending for 38,600 sq. miles (100,000 km²) in Venezuela, and considerably farther in adjoining Colombia, the Llanos are smoky with frequent forest fires in the dry season, when the wildlife takes refuge in the valley bottoms and along the river banks. Winter rains create a huge marshy territory of rivers and streams, lakes and swamps.

The spectacled caiman, reduced in numbers by poachers, is now making a comeback in the Llanos. The waters are also perilous with red piranhas, electric eels and stingrays. The grasses and scattered palms of the flood plain and the trees of the riverine woodlands provide cover for many creatures, from capybara, white-tailed deer and jaguars, to tree frogs and iguanas. Internationally important as a wading-bird refuge, the wetlands host large breeding colonies of many species of heron, stork and ibis.

27 Maraca BRAZIL

A huge riverine island, Maraca lies some 75 miles (120 km) northwest of Boa Vista, in the northern state of Roraima. Covering an area of about 170 sq. miles (440 km²), Maraca is uninhabited rainforest terrain, with patches of open savanna, seasonally flooded wetlands, creeks and low hills.

Maraca teems with life and represents the forests of the Amazon basin in microcosm, secluded and unspoilt. There are 450 species of birds and almost 50 species of bats. This is no tourist spot. Large and aggressive packs of peccaries patrol the forest, and the nests of hornets and killer bees hang from branches. Jacaré alligators lie submerged in the ponds, and the encircling rivers are home to piranhas, as well as stingrays. Rattlesnakes, boa constrictors and deadly fer-de-lance snakes are common. Mammal life includes jaguars, three-toed sloths, several monkey species, great anteaters, agouti, tree porcupines, tapirs and deer. A Royal Geographical Society project in Maraca has discovered hundreds of species new to science.

28 Chiribiquete National Park COLOMBIA

The park is an outstanding example of sustainable indigenous agriculture at work. Situated on the western edge of the tropical rainforest, against the foothills of the Andes, Chiribiquete covers 3,860 sq. miles (10,000 km²) and is part of the transfer of some traditional forest lands back to Indian ownership. The Colombian government has made the transfer in recognition of the fact that the Indians' agricultural methods are environmentally sound and efficient.

The Indians exist in fairly small family groups, hunting and fishing, and foraging through a wide area for wild produce. In addition, they clear forest gardens, away from the rivers, in which they plant crops including mangoes, papayas, yuccas and peppers. Like most artificially cultivated rainforest soils, these gardens begin to lose their fertility after a couple of years. The Indians then abandon the old gardens and move on to new ones. Over a period of a number of years, the forest gradually recolonizes the area in the same way that it does when a fallen tree creates a clearing.

29 Galápagos Islands PACIFIC OCEAN

The islands, islets, rocks and reefs of the Galápagos archipelago, some 500 miles (800 km) west of the Ecuadorian coast, were made famous by Charles Darwin's visit and subsequent writings. The larger islands, including Isabela with over half of the group's land area, have arid coastal lowlands, dry from low precipitation and dark with volcanic soil. Born of volcanic action only a few million years ago, the islands still have several active peaks. Vegetation ranges from cactuses on the coast to a zone of evergreen forest between 660 feet (200 m) and 1,650 feet (500 m), with ferns and sedges above that.

The Galápagos' flora and fauna must have been conveyed by wind and sea currents, travelling as seeds and eggs on natural vegetation rafts or on the feet and feathers of seabirds. The unique fauna includes giant tortoises, the planet's only marine iguanas, Galápagos fur seals and sea lions, two bat species, many seabirds and many landbirds, including the 13 finch species showing distinct local adaptations, which influenced Darwin's thinking on evolution.

30 Pacaya Samiria National Reserve PERU

The mighty Amazon rises in the Peruvian Andes, and the Pacaya Samiria National Reserve is situated on the Amazon's head-waters, on and around the rivers Pacaya and Samiria and 5,330 sq. miles (13,800 km²) of mixed waterways, oxbow lakes, lagoons, islands and forest. Initially established as a river reserve to protect the paiche, or giant catfish, endangered by overfishing, the enlarged territory is now a refuge for all species of fauna and flora, with a management plan to allow limited and sustainable harvesting by locals for their own needs.

Apart from the huge paiche, stingrays, piranhas and many other fish species, the waters of the reserve contain Amazonian dolphins and caymans. A great variety of creatures inhabits the forests lining the rivers, including wooly and squirrel monkeys, deer, peccaries and two large rodents, the capybara and the paca, both of which are among the sustainable game populations subject to limited hunting.

31 Tambopata-Candamo Reserve PERU

The Peruvian government established this 5,800-sq. mile (15,000-km²) region as an "extractive reserve" for forest products such as rubber and brazil nuts, which can be harvested without harming the ecology of the area and help pay for its management. The reserve stretches from the High Andes down into low jungle, and encompasses the entire watershed of the Tambopata River. Organized ecotourism also contributes to Tambopata's self-sufficiency.

Humans have to share brazil nuts in the reserve with macaws, which have evolved enormously strong beaks to open them. The agouti, a large rodent, also eats the nuts and helps the spread of the plant by burying clusters of uneaten nuts.

32 Lake Titicaca PERU

Legendary birthplace of the Inca nation, at 3,800 m (12,500 feet) high in the Peruvian Andes and almost 3,200 sq. miles (8,300 km²) in area, Lake Titicaca is the world's highest navigable lake. It is fed by snow melt from the surrounding peaks, and the depth varies by 3 feet (1 m) between the wet and dry seasons. Titicaca has only one outlet, the Desaguadero River,

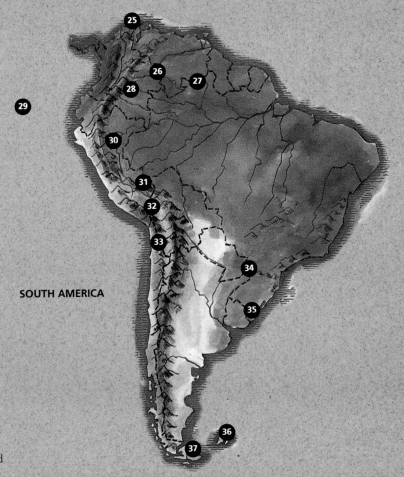

SOUTH AMERICA

which takes about 10 percent of all water leaving the lake. The rest disappears as a result of evaporation, aided by high altitude, strong sunshine and the constant winds of the Altiplano.

In the shallows of the lake grow large areas of the long totora reeds used by the local Indians to build their boats and floating houses. Such a huge area of water, with extensive shallows, is an important winter stopover for migrating North American shorebirds, as well as a permanent home to high Andes waterbird species. Perhaps the most unusual lake dwellers are the frogs, which live on the lake bottom, can breathe through their skin and seldom surface.

33 Atacama-Sechura Desert CHILE

The 1,865-mile (3,000-km) ribbon of the Atacama-Sechura Desert forms a narrow coastal strip overlooked by the massive heights of the Andes. It is one of the most desolate and arid places on Earth, with some regions having a barely measurable 1 mm of precipitation a year. Yet, despite this lack of rain, caused by the rain shadow cast by the Andes, relative humidity can be high, with long periods of fog and cloud. The bleakness is emphasized by the cold, as onshore winds blow in over the icy Humboldt (Peru) Current.

The Humboldt penguins that swim in the cold current are one of the very few forms of life visible in this grim landscape, nesting incongruously under cactus plants at the ocean's edge. In the most arid section of the Atacama, the temperature can drop at sunset from 104°F (40°C) to 32°F (0°C) in an hour. There are a few small rodents and lizards, and the foxes and birds of prey that hunt them, but the overriding character of the Atacama is barrenness, in stark contrast to the rich wildlife on the other side of the mountains.

34 Iguaçu Falls BRAZIL

The great cascade of the Iguaçu Falls pours over the fractured rim of Brazil's Parana Plateau through a sieve of islands covered in vegetation, creating a rank of almost 275 separate falls. The rim over which the wide river plunges is 2½ miles (4 km) in length, and the gorge funneling the waters downstream lies 269 feet (82 m) below the rim. The whole lush region around the falls has long been protected parkland, established by the governments of Argentina and Brazil on their respective sides of the border.

The Iguaçu is South America's third largest river, yet on average it dries up completely every 40 years or so, silencing the cascades for up to a month. Most of the time, the falls thunder over their cliffs of volcanic basalt, in full spate carrying seven times as much water as Niagara and filling the air with a constant mist, encouraging the growth of ferns, mosses and unusual aquatic herbs, which use suckers to cling to the wet rocks.

35 Bañados del Este BRAZIL/URUGUAY

A broad barrier of dunes separates the wetlands of the Brazil-Uruguay coastal border from the Atlantic Ocean, protecting the large, brackish Laguna Merim and its associated freshwater marshes, peat swamps and flood lands. Great flocks of migrating birds breed in these shallow waters, including those coming from farther south during the southern winter and from North America during the southern summer. The entire complex of bodies of water, seasonally flooded grassland and palm savanna covers an area of 4,600 sq. miles (12,000 km²).

Local communities used to harvest coypu in the Bañados for their fur, but the main threat to the area today comes from agriculture and pressure to increase the number of drained areas. Wetlands on both sides of the border have been designated reserves, and numerous coypu still live in the marshes, as well as capybaras.

36 Falkland Islands SOUTH ATLANTIC

Lying 280 miles (450 km) off the Argentine coast, the Falkland Islands are low and windswept, subject to raw westerly gales that keep the vegetation down to shrubs and tussock grass. There are two main islands and numerous offshore islands, and the coasts are rocky and deeply indented with bays and inlets, with cliffs in some places. Some 53 of the offshore islets have been designated reserves.

Due to the lack of trees, the vast numbers of seabirds that come to the Falklands to breed have to nest on or close to the ground. They include shearwaters, black-browed albatrosses and four species of penguins. The bays and inlets of the islands attract populations of southern sea lions, elephant seals and fur seals, all of which were severely depleted in the 19th century, but have now recovered substantially. The most unusual mammal is the sea otter, small numbers of which live among the thick kelp beds around offshore islands.

37 Tierra del Fuego CHILE

The archipelago of Tierra del Fuego is a harsh wilderness of mountainous islands and ragged channels, sculpted by retreating ice and still being eroded by howling westerly gales. Annual rainfall on its western edge is 16 feet (5 m). A graveyard for ships in the days of sail, these inhospitable rocky shores are sometimes fringed with beech forests, although most growth is found on the eastern sides of the islands, out of the icy winds that gust into squalls of 115 mph (100 knots).

The fauna of the region is concentrated in the waters, where the teeming krill supports packed food chains of fish, seabirds, southern sea lions, dolphins and whales. The shoreline rocks are festooned with giant kelp, and the abundant shellfish were once the staple diet of the Yahgan Indians who lived here until the arrival of Europeans caused them all to die through measles. In the eastern tundra-covered plains of the main island of Tierra del Fuego, ducks and geese breed. On the east coast are tidal mud flats used as temporary habitats by birds such as Hudsonian godwits and white-rumped sandpipers which have migrated from Arctic breeding grounds.

AFRICA

38 Banc d'Arguin National Park
MAURITANIA

The Banc d'Arguin is a huge Atlantic coast intertidal wetland system of mud flats, creeks and eelgrass beds. The 3,860-sq. mile (10,000-km²) national park includes the Baie d'Arguin, and the region is a frontier zone between systems, where the Sahara Desert meets the Atlantic Ocean, and the cold Canaries Current from the north meets the much warmer Guinean Current from the south.

The local Imraguen tribespeople fish for migrating mullet in the shallow waters by beating the water with long poles to attract dolphins, which come and drive the fish into the fishermen's nets. The Banc d'Arguin's rich diversity of fauna includes several million migrating shorebirds, including dunlin and bar-tailed godwits, as well as breeding colonies of spoonbills, greater flamingos and great white pelicans. The largest population in the world of the endangered Mediterranean monk seal lives in the park's waters, and marine turtles lay their eggs on the beaches under the protection of camel-mounted wardens.

39 Air and Ténéré National Nature Reserve
NIGER

Covering over 29,730 sq. miles (77,000 km²), or 6 percent of Niger's total area, the Air and Ténéré National Nature Reserve is a hot and arid wilderness of desert and mountain that is a last refuge for many of the creatures of the Sahara and the Sahel that have virtually disappeared elsewhere. It is one of Africa's largest conservation areas, with a small human population, mainly Tuareg nomadic pastoralists.

In the desert portion of the reserve, there are furnacelike areas of shifting dunes where no plants grow, and other areas with a sprinkling of hardy acacia trees. In the Air Mountains, humid mists condense in the cold nights into groundwater, which lies in shallow pools and encourages the growth of palms, as well as Mediterranean-type trees such as wild olives. With its combination of desert and mountain terrains, the reserve contains significant populations of creatures such as ostriches, oryxes, Barbary sheep, gazelles and cheetahs, as well as the endangered Saharan antelope, the addax.

40 Bijagos Archipelago GUINEA BISSAU

The 88 islands and islets of the Bijagos Archipelago cover an area of 3,860 sq. miles (10,000 km²) in an ancient river delta. A tangled web of creeks fringed with mangroves drains the wide mud flats at low tide. Great volumes of fresh water drain down to the sea in the rainy season, and tropical forest comes down to the beach on some islands, interspersed with the palm groves from which the indigenous Bijagos people extract oil and palm wine.

The shallows and mangroves are rich with shellfish, harvested by human islanders, migratory shorebirds, and the otters that swim and fish in and around the creeks. Hippopotamuses in Bijagos have adapted to a beach and saltwater existence, and the tidal creek waters provide a refuge for a large manatee population. Other creatures benefiting from this extensive maze of islands and tidal waterways are four species of marine turtles, two species of crocodiles and two species of dolphins, including the rare Guinean dolphin.

41 Gashaka Gumpti Reserve NIGERIA

Situated close to Nigeria's border with Cameroon, the Gashaka Gumpti Reserve lies in a remote region that combines a rich mixture of uplands, lush rainforest and savanna grasslands. Established in a cooperative effort between central and local government and nearby villages, the management of the reserve maintains a successful balance between wildlife protection and local human demands.

The huge diversity of wildlife in the reserve includes anubis baboons, Ethiopian colobus monkeys and at least six other primates, with the possibility of mountain gorillas in mountain areas so remote they have never been explored. There are leopards and rare golden cats in the rainforest, and plains species include wildebeest, hartebeest, antelopes and gazelles. This abundance of prey animals encourages the leopards that enjoy the reserve's protection.

42 Sudd Swamp and Flood Plain SUDAN

Occupying up to 12,400 sq. miles (32,000 km²) of the basin of the Upper Nile in central Sudan, the Sudd is a humid, rain-drenched wetland where large floating mats of papyrus move slowly on the shallow currents, blocking river channels and spreading the great river's headwaters over a wide area. Once a year, at flood time, the Sudd doubles in area, forcing animals and people to move temporarily to higher ground or into surrounding regions.

Here are some of the biggest gatherings of large mammals left in Africa, in the form of the periodic mass migrations of tiang and white-eared kob antelopes. Here too is the Nile lechwe, an antelope specifically adapted to wetland existence, with elongated hooves that enable it to move through the swamp without sinking. Hippos, crocodiles, giraffes and elephants are all found in the Sudd, which is also Africa's richest wetland bird territory. Threatened by proposed agricultural expansion, the Sudd is of vital importance to the entire Nile ecosystem, as well as being a major wildlife sanctuary.

43 Abijatta-Shalla Lakes National Park
ETHIOPIA

East Africa's major portion of the Great Rift Valley, running from the Red Sea in the north to Lake Manyara, Tanzania, in the south, threads a necklace of lakes, and Ethiopia's Lakes Shalla and Abijatta are among those not too tainted with sodium carbonate to support a varied wildlife.

Separated by a 8,860-foot (2,700-m) mountain ridge, the lakes are very dissimilar in character. Shalla has a depth of 855 feet (260 m), contains the most important island nesting colonies of great white pelicans in the world, but is too deep and steep sided to provide spawning grounds for fish. As a result, the adult pelicans soar up on thermal currents to cross the mountain ridge and go fishing in the shallow waters of Lake Abijatta, returning 24 hours later on the following morning's thermals to feed their young. Hosts of birds, including herons, ibises, kingfishers, cormorants and darters as well as the commuting pelicans, frequent Abijatta's favorite fishing grounds to feast on tilapia and other fish.

44 Ruwenzori Mountains ZAIRE/UGANDA
The upper slopes of the Ruwenzori Mountains are mysterious with mists, festoons of lichen and

gigantic flowering plants. In the zone between approximately 10,830 and 13,125 feet (3,300 and 4,000 m), there are few animals, but plants known as small species in temperate climates exist in gigantic forms specially adapted to survive the extremes of climate. Lobelia and groundsel reach heights of up to 20 feet (6 m), and heathers can grow to tree-size of 40 feet (12 m) and more, stimulated by constant moisture, lack of tree competition, mineral-rich soil, and plenty of ultraviolet radiation.

Between 4,920 and 7,875 feet (1,500 and 2,400 m), the grass-covered foothills of the range change to a canopied rainforest of cedar, camphor and podocarp. Above that is a dense belt of mountain bamboo, superseded at 10,000 feet (3,000 m) by subalpine moorland. Giant flowering plants inhabit the next zone, and above that are the permanent snowfields and glaciers found only here and on neighboring Mounts Kilimanjaro and Kenya.

45 Ngorongoro Crater TANZANIA
Some 2.5 million years ago, a volcano collapsed into a drained subterranean magma chamber forming a caldera, which today is known as the Ngorongoro Crater. The floor of the crater measures 100 sq. miles (260 km²) in area, and is surrounded by a rim some 2,000 feet (610 m) high, making Ngorongoro into a vast natural amphitheater with a floor of savanna grassland. A shallow soda lake lies at the crater's lowest point, where teeming flamingo flocks of both lesser and greater species sift for food.

The animals of Ngorongoro are not prisoners of the crater, and some do leave by old trails over the rim. However, there is little incentive to emigrate, since food is usually plentiful and there is a constant water supply. Elephants and hippos frequent the river-fed swamp in the north of the crater, joined in the dry season by gazelles, zebras and wildebeest. Buffaloes, warthogs and ostriches live in the crater, and taking advantage of the plentiful supply of prey animals is a variety of predators, including lions, cheetahs, hyenas, jackals and foxes.

46 Mount Kilimanjaro
TANZANIA

Africa's tallest mountain has a permanent cap of snow and ice despite its proximity to the equator. It has three peaks, and the tallest, Kibo, at 19,341 feet (5,895 m), has a neat circular crater 1½ miles (2.5 km) across. All three peaks are dormant volcanoes, and Kibo may last have erupted within the past few hundred years.

The base of the great mountain is huge, 50 miles (80 km) by 25 miles (40 km), and between 6,235 and 9,845 feet (1,900 and 3,000 m) it is encircled with a zone of thick montane forest. Into this forest of liana-festooned cedars and buttress-rooted podocarp trees come

elephants and Cape buffaloes, bushbuck and duiker, but they are relatively few and far between, and keep quiet and out of sight. Above the forest zone are moorlands of heather, tussock grass and giant groundsel and lobelia. The rocky slopes and valleys above the moors are rough with porous lava. Leopards roam these bleak upper levels, as well as eland, and the strange little rock hyrax, cousin to the elephant, shrieks piercingly in the still air.

47 Etosha National Park NAMIBIA
For most of the year the 2,320 sq. miles (6,000 km²) of the Etosha Pan in the Etosha National Park is dry, and the sun glares off the surface crust of salt. Flamingos and pelicans gather on the highly saline shallows that follow sporadic rain, but beneath the Pan lies a large enough natural reservoir to feed the surrounding pools and springs that deliver water year round.

From all over the greater park area beyond the Pan, the animals home in on the waterholes in the dry season. Magnetized by the scent of water, the herds gather: zebra, giraffe, kudu, gemsbok, wildebeest, hartebeest and elephant. Every so often, summer rains are torrential. Dormant seeds germinate, and the brilliant green of grass and other plant growth covers the clay. The herbivore herds trek off toward the rain and new shoots. Lion families are forced to live on lesser game and birds until the herds return in the winter. Constantly poised in a delicate and perilous balance between survival and death, Etosha remains one of Africa's most magical refuge wildernesses.

48 Madagascar INDIAN OCEAN
Lying 500 miles (800 km) from the coast of southeastern Africa, Madagascar is the fourth largest island in the world, 976 miles (1,570 km) long by 354 miles (569 km) across at the widest point. Hilly woods in the west of the island lead to a central plateau, beyond which, to the east, is a tropical rainforest that catches the rains blowing in from the ocean to the east. In the south there is an arid desert region.

Madagascar is a rare repository of strange and unique creatures, which have developed there in isolation since the island split away from the African mainland 100 million years ago. All the planet's lemurs and half its chameleon species live on the island. Other unique Madagascan creatures include almost 150 species of frogs and 30 species of the unusual, hedgehog-like tenrec family. Many of the island's birds originate from the African mainland, but there are over 40 unique birds, such as the cuckoo roller and the chameleon-eating Madagascan serpent eagle. The island is particularly rich in plants, with an estimated 9,000 species.

49 Kalahari Desert BOTSWANA
With five months of wet summer from October to March, the Kalahari Desert actually receives a much higher average rainfall than a true desert. More rain falls in the north than the south, reflected in vegetation growth: woodland in the north, with ebony, sycamore and baobab; scrubby grassland in the south, with occasional clumps of palms. The rainfall of this semidesert means a wide range of wildlife lives throughout the 100,400-sq. mile (260,000-km²) plateau of the Botswana Kalahari.

The rains form temporary streams and water pans in the depressions between sandy ridges. Because of flooding in the Okavango Delta in the north, many animals migrate south into the temporarily wet desert regions, including springbok, elephants, zebras, wildebeest and buffaloes.

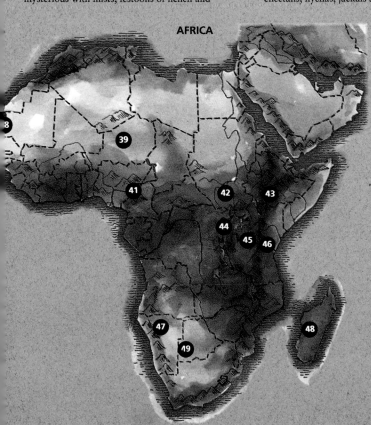

AFRICA

EUROPE

50 Hornstrandir Landscape Reserve
ICELAND

Wild even for Iceland, Hornstrandir is a deeply weathered headland of cliffs and rocky peninsulas in the country's far northwest. Since it is close to the Arctic Circle, the reserve suffers ferocious winters, with heavy, long-lasting snow and winter drift ice. Summer rains and fogs can also turn into snow storms, and visitors have to approach the 225-sq. mile (580-km²) region either by sea, or on foot over pathless terrain. Hornstrandir is truly one of Europe's most rugged wildernesses, yet it repays the persistent visitor with magnificent scenery and fascinating wildlife.

Ancient lava sheets cover the remains of even more ancient forests, but today Hornstrandir is brilliant with wildflowers in spring and summer, at the same time that the cliffs and ravines of the reserve are acting as nurseries to teeming masses of seabirds. Fulmars, kittiwakes, guillemots, puffins, razorbills and gulls cram every spare inch of space on the ledges, while white-tailed sea eagles and gyrfalcons patrol the bird-cliffs in search of prey.

51 Linnansaari FINLAND

A third of Finland consists of lakes and wetlands, and the small national park of Linnansaari occupies just 3 sq. miles (8 km²) of water and small islands in the large Lake Saimaa, in the country's southeastern lake district. There are many islets in the park, some just rocks, but most thickly wooded with conifers such as spruce and pine as well as deciduous trees like lime, birch and alder. Despite its diminutive size, Linnansaari is a collection of self-contained worlds, with deep inlets and wooded glens, and a lush ground cover of shrubs such as bearberry, bilberry and cowberry.

The lake waters attract goosander, mallards, gulls and black-throated divers, and ospreys breed within the park. The largest island is only 2½ miles (4 km) long. Elk are the largest of the forest fauna, but the most celebrated local mammal is a rare subspecies of Saimaa seal, which may originally have reached the lake from the sea before it was landlocked during the last ice age. It is now one of the world's very few freshwater seals.

52 Hardangervidda NORWAY

Norway's Hardangervidda National Park, in the south of the country, is a harsh, mountainous plateau where few trees grow. The park covers 3,860 sq. miles (10,000 km²) of rocky peaks and icy valleys between 3,935 and 5,575 feet (1,200 and 1,700 m) in altitude, where snow and ice linger on the uplands and on the multitude of lakes for much of the year. The largest herd of wild reindeer in Europe, some 1,400 strong, subsists within the Hardangervidda borders on a diet of mosses, lichens and the dwarf plants of the windy uplands. The region has two climates: a coastal climate in the west where it comes close to the coastal fiords, and a colder inland climate farther east. The region is rich in arctic and subarctic plants, and supports a bird population typifying both arctic and temperate zones, with dotterel and great snipe as well as plovers, wagtails and warblers.

53 Bayerische Wald National Park
GERMANY/CZECHOSLOVAKIA

At more than 50 sq. miles (130 km²), with 98 percent tree cover, this is central Europe's largest forest region, a mountainous treescape, rich in plant species, with an abundance of streams, marshes and boggy valleys. A well-organized logging center in the last century, the Bayerische Wald is now a forest in the process of returning to the wild. Streams are re-establishing their ancient courses after having been redirected to act as lumber channels. The drainage ditches dug to facilitate access are now choked with organic debris and vanishing as the forest reclaims its natural form.

Along the Czech border, the park is corrugated with mountain ridges. Spruce is the most common tree at the higher forested levels, merging into mixed colonies of fir, ash, maple, elm, alder, willow and bird cherry farther down. A profusion of woodland and marsh flowers and mosses grows beneath the tree cover and in the meadow clearings and marshes. The rich birdlife includes honey buzzards, pygmy owls and four species of woodpeckers.

54 Höhe Tauern AUSTRIA

The Grossglockner, the loftiest peak in the Austrian Alps at 12,457 feet (3,797 m), dominates the center of Höhe Tauern, Austria's major protected region. Lying in an alpine region of granite, gneiss and schist, Höhe Tauern contains a number of peaks over 10,000 feet (3,000 m). The park has a wealth of breathtaking views and features, from rocky summits streaked with snowfields to thundering waterfalls hundreds of yards high. The Krimml Falls, in the northwest of the park, descend 1,300 feet (400 m) in three great cascades.

With an area of 965 sq. miles (2,500 km²), Höhe Tauern is larger than all the other Austrian reserves put together. There are extensive alpine woodlands along the lower valleys, but the most spectacular wild inhabitants are the great soaring birds of prey which can fully exploit the peaks, cliffs and ravines of the mountain ridges. Golden eagles, griffon vultures and even the rare lammergeier are all visitors to the park.

55 Swiss National Park SWITZERLAND

Situated in the far eastern mountains of the Engadine, adjoining the Italian border, the Swiss National Park was established in 1914 and has a reputation as the best-managed conservation area in Europe. Most of the park lies above 6,500 feet (2,000 m), and it has three major zonal terrains. Evergreen forests, reaching up to a tree line of some 7,550 feet (2,300 m), give way to high alpine meadows, covered in flowers in early summer. Above the meadows are the screes and limestone outcrops of the high peaks and ridges.

The larger carnivores, such as bears, wolves and lynxes, became extinct by the early years of the 20th century, as they did throughout most of the Alps. Ibex and red deer also disappeared, but have returned. Marmots and chamois are common.

56 Saja National Reserve SPAIN

Close to Spain's northern Atlantic coast, in Santander province, the Saja Reserve is named after the Saja River, which is one of several watercourses cutting northward across it to drain eventually into the Atlantic. The reserve also contains the source of the great Ebro River, which runs southeast to enter the Mediterranean. The limestone ridges of the Cordillera Cantabrica in the reserve were once flat plains, but are now tilted on edge. Beech trees have a natural affinity with limestone, and the forests covering most of the reserve are mainly of beech and oak. Some of the trees reach an enormous size, with beeches up to 100 feet (30 m) in height.

Although there are a number of villages within the reserve, the terrains of hill forest, wooded river valley and marshy wetland harbor an exciting range of wildlife, including species that have become extinct elsewhere. Brown bears and wolves survive here, as do populations of boar, wildcat, chamois and roe deer.

57 Pyrenees National Park FRANCE

Stretching along the border with Spain, France's Pyrenees National Park is part of a rugged natural barrier between the two countries. Rocky peaks and grassy uplands predominate, but the park's features also include lakes, waterfalls, glacial basins and thickly forested slopes and valleys. The tree line here is the highest in Europe, with individual trees surviving at altitudes up to 8,530 feet (2,600 m). Mountain pines at high altitude often shelter a dense shrub layer of flowering alpenrose, a species of wild rhododendron. Beechwoods line many of the valleys. In the south and east of the park are the large glacial amphitheaters of the Cirque de Gavarnie and the Cirque de Troumouse. Cirques are ice-worn basins, and the Pyrenean cirques feature cliffs, waterfalls, lakes and flower-filled grass floors.The park is particularly rich in mammal wildlife, with boars, genets, chamois and, rarest of all, brown bears.

58 Ordesa National Park SPAIN

Adjoining a particularly remote sector of the French border which it shares with France's Pyrenees National Park, Spain's Ordesa National Park occupies one of the country's least-known regions. The steep-sided and forested Ordesa Valley formed most of the park's territory from 1918 until 1978, when three further valleys were added, bringing the total area up to 61 sq. miles (157 km²). The highest point in the park is Monte Perdido, at 11,007 feet (3,355 m). The valleys of Ordesa form some of Spain's most isolated environments. Spectacular limestone rock faces and cliffs tower above the forests, and there are many ravines and waterfalls, as well as glaciers and high mountain meadows.

The forests are mainly of Scots pine, mountain pine, beech and silver fir, and their wild inhabitants include otters, wild boars, polecats, Pyrenean ibex and wildcats. The strange, molelike Pyrenean desman hunts in the streams, living in burrows in the banks.

59 Lake Mikri Prespa GREECE

High in the remote mountain borderlands that divide the territories of Greece, Albania and the former Yugoslavia, the shallow waters of Lake Mikri Prespa host one of Europe's most important waterbird populations. The lake is surrounded by extensive reed swamps and marshy lagoons, all forming part of the national park set up by the Greek government in 1971. Mikri Prespa and the adjoining Lake Megali Prespa lie in a bowl-like depression, with a backdrop of gray Albanian mountain peaks. Every spring, torrents of fresh ice-melt recharge the lake waters and flood the surrounding water meadows.

Large colonies of waterbirds breed in the marshes, among them Dalmatian pelicans and white pelicans, both species very rare in Europe. Other waterbirds breeding around the lake include egrets, cormorants, herons, spoonbills, ibises and bitterns. Among the mammals surviving in this inaccessible region are brown bears, wolves, jackals and otters.

60 The Vikos Gorge GREECE

The great chasm of the Vikos Gorge, with its exhilarating views, limestone cliffs and precipitous side canyons, is less well-known than Crete's Samaria Gorge, but is equally impressive, and considerably more difficult to explore. Situated in the mountain territory close to the Albanian border, the gorge contains the source and the upper course of the Voidhomatis River. The icy spring water of

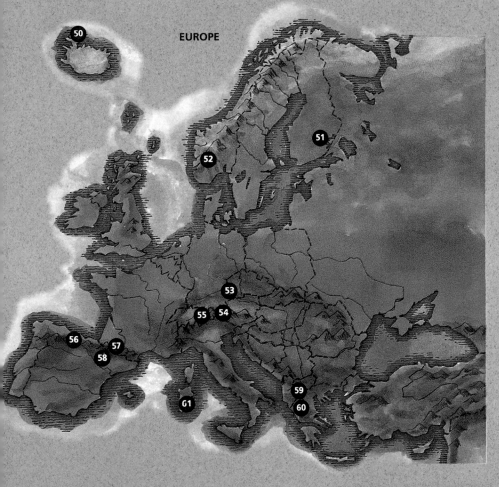

EUROPE

ASIA

the source bubbles out from the base of astonishing cliffs, soaring upward well over 4,900 feet (1,500 m). The total length of the gorge is a little more than 8 miles (13 km).

Dense coppices of maple, beech and chestnut flank the trail in some upper stretches of the gorge, and where it is joined by the Megas Lakkos Gorge, the Klima spring creates a damp and shadowed environment of ferns, moss and slippery boulders. The region around the gorge still harbors bears, wolves and the European jackal, while golden eagles, Bonelli's eagles, griffon vultures and peregrine falcons haunt the tall cliffs.

61 Gennargentu SARDINIA, ITALY

Sardinia lies 110 miles (180 km) west of the Italian mainland. Much of the island still consists of wild and difficult country, and the Gennargentu National Park, on the eastern coast, occupies some 390 sq. miles (1,000 km²) of Sardinia's wildest terrain. The Gennargentu massif, with several peaks more than 4,900 feet (1,500 m) high, forms the heart of the park, with forest-covered ridges and valleys, and, in the east, dramatic ravines and gorges descending to the Gulf of Orosei.

Large tracts of the Gennargentu territory is made up of maquis, the typical scrubland of the Mediterranean coastlands, consisting of tough shrubs like the strawberry tree and the tree heath, often several yards in height, which are resistant to drought and have replaced the pine and holm-oak forests cleared centuries ago. Maquis is an ideal environment for a variety of creatures, including lizards, snakes, spiders, scorpions and praying mantises. Gennargentu's most celebrated wildlife is its protected mouflon population, which lives within a special reserve in the park.

62 Kamchatka EASTERN SIBERIA

Hanging southward for 750 miles (1,200 km) from the eastern end of Siberia, the Kamchatka Peninsula reaches out into the Sea of Okhotsk toward the islands of Japan. Sparsely inhabited, and gloomy with sea fogs much of the time, Kamchatka seems to live in perpetual winter, enlivened by 33 active volcanoes along its length, which add their steam to the low cloud cover. Three of these volcanoes are more than 13,000 feet (4,000 m) high, and one, Klyuchevskaya Sopka, has erupted more than 70 times since 1697, most recently in 1990. Kamchatka contains 400 glaciers and a number of hot springs and geysers.

Vegetation along the rivers flourishes in warm, mineral-rich soil. Each year, salmon migrate up to breeding grounds in the Kamchatka rivers, preyed on by harbor seals along the coast, then by humans. The huge brown bears of the peninsula devour dead and dying fish, exhausted by their migration journey, in preference to catching them, scavenging alongside northern ravens, slaty-backed gulls and the magnificent Steller's sea eagle.

63 Ussuriland SOUTHEASTERN SIBERIA

Plant and animal species of the north meet those from Southeast Asia in Ussuriland, one of Siberia's most diverse wildlife areas. The Sikhote-Alin mountain range, home to the Siberian tiger, runs from north to south for 620 miles (1,000 km) parallel to the coast, with the meadows and wetlands of the Khanka lowland to the southwest.

The northern birch, spruce and pine woods of Ussuriland sustain large populations of birds such as woodpeckers and nutcrackers. Lynxes and wolverines hunt through the trees, which are also the

territory of boars, brown bears and several deer species. At the nature reserve of Kedrovaya Pad, close to the Korean and Chinese borders, the southern character predominates. The forest is brightened and perfumed by flowering shrubs and summer flowers, while large and gorgeous butterflies such as the Maack's swallowtail fly between the trees. Kedrovaya Pad is also a refuge for a few pairs of the extremely rare Amur leopard.

64 Shiretoko National Park JAPAN

Situated on a peninsula at the northeastern end of Hokkaido, Japan's remote, northernmost island, Shiretoko is the country's wildest park. For more than six months of the year, Hokkaido is covered in snow, and winters can be so cold that the inshore waters fill with ice floes, and the rivers and waterfalls that plunge over the peninsula's cliffs freeze solid. During the bitter winter, many seabirds move inland, and at sea the Steller's sea lions move south if inshore ice makes fishing too difficult.

The Shiretoko Peninsula has a spine of volcanic peaks, one of which is still active. Below the volcanic ridge, thick forests encircle the Shiretoko Five Lakes, which are situated in the underlying lava bed. These forests are home to brown bears, foxes and deer. Forestry Bureau plans to log the reserve threatened the habitats of the rare Blakiston's fish owl, a totemic creature to the indigenous Ainu people. Public outcry from all over the country has led to an indefinite postponement of the plan.

65 Astrakhan Reserve VOLGA DELTA, CASPIAN SEA

Lenin created this, the former Soviet Union's first reserve, in 1918, to preserve the Volga Delta from overfishing and bird poachers. Situated at the northern end of the Caspian Sea, the delta is a wildlife oasis set in an enormous area of relatively dry plains and grasslands. Over 250 species of birds use the reserve, many of them fish-eaters. Among the 60 or so mammal species is the Caspian seal.

The channels and streams of the delta are lined with willows, tamarisks and reed beds. The enormous variety of birds feed on the abundant insect, amphibian and fish life. Herons spear frogs, terns trawl the air for insects, and cormorants dive for fish. In the shallows, wild boars dig in the mud with their snouts for roots and snails. There are 100,000 great cormorants in the Astrakhan Reserve, in colonies of 1,000 and more pairs. They build nests of sticks more than 3 feet (1 m) across and fly great distances to catch fish for their young.

66 Tien Shan CHINA/KYRGYZSTAN/KAZAKHSTAN

In Chinese, Tien Shan means "Celestial Mountains," and these border mountains between China and the independent republics of the former Soviet Union have a number of summits over 16,400 feet (5,000 m). The upper reaches of the Tien Shan are covered in perpetual snow and ice, with many glaciers. The foothill plains can be baking hot, but the temperature of the air drops 4°F (2°C) for every 1,000 feet (305 m) of altitude, and the lower slopes are forested with spruce and junipers. Above 9,300 feet (2,800 m) the cold is too severe for forest growth, and juniper scrub eventually gives way to alpine meadow and scree.

Lammergeiers, Himalayan griffons and Eurasian black vultures all ride the air currents of the high ridges in search of carrion. Spring on the southern slopes brings a thaw and the emergence of brilliant bulb flowers, including juno irises and many species of wild tulip, followed by lilies, delphiniums and aconites. Mammal life includes Marco Polo sheep,

Siberian ibex and wild boars, which are found as high as 10,000 feet (3,000 m). Hibernating marmots emerge in the spring, as do brown bears, many of which have very pale fur and white claws.

67 Repetek Biosphere National Reserve
TURKMENISTAN

The Repetek Reserve lies toward the arid edge of the Karakum desert region in southeastern Turkmenistan. The whole area is a stifling, wind-scoured wasteland of sands, dunes and dry clay. At first sight, the land seems lifeless, but on closer inspection it turns out to harbor a thriving variety of desert-adapted flora and fauna. The reserve, covering about 135 sq. miles (350 km²), is a microcosm of the Karakum.

Among the shifting barchan dunes are groups of black saxaul, a deep-rooted and leafless desert tree which grows in groves, usually in low ravines. Camelthorn, anchored with 100-foot (30-m) tap roots, studs the ridges, and gopherlike susliks inhabit complex colonies in the slopes, preyed on by foxes, jackals and sand cats. Best adapted to the hot sand are the spiders, scorpions, insects and reptiles, including tortoises, monitor lizards, toad-head lizards, and a number of species of snakes – some, like the blunt-nosed and saw-scaled vipers, very poisonous.

68 Dachigam National Park INDIA

Beginning at Marsan Lake, the Daghwan River flows through the Dachigam National Park toward Srinagar, collecting tributaries as it goes. The whole park, tucked up tight against India's northern border, is a catchment area for Srinagar's drinking-water supply. Dachigam is only 54 sq. miles (140 km²) in area, but it manages to incorporate most of the habitats of the fertile and mountainous Himalayan approaches within its borders, from riverine forests to bare rock and alpine scrub.

In the winter, the snow lies thick, and many high-altitude birds move down to where food is available. Endangered hangul deer, for which the park is the world's major refuge, move to the shelter of the lower valleys. In spring, the Himalayan black bears come out of hibernation, and the short summer sees the alpine meadows ablaze with alpine flowers.

69 Arjin Shan Reserve, CHINA

In addition to being one of the world's largest nature reserves, at 17,375 sq. miles (45,000 km²), Arjin Shan is also one of the last of the real wildernesses – remote, uninhabited and virtually inaccessible. The reserve consists of an enormous relatively flat plain, containing large freshwater and saline lakes, and encircled by a towering ring of snow-covered peaks. The lowest part of the reserve is at 10,170 feet (3,100 m), and its highest point is the peak of Wu-lu-k'o-mu-shih at 25,338 feet (7,723 m).

Arjin Shan is one of the last places left in Asia where large herds of hoofed mammals can travel unimpeded across the great areas necessary for them to find sufficient grazing. Herds of Tibetan antelope and wild ass move across the plain, congregating on the shores of the lakes when there is enough grass to crop. In the mountains are the herds of hardy high altitude feeders, like wild yak and Tibetan gazelle. Higher up still, among the rocks of the high meadows, forage ibex and blue sheep. Naturally, predators are also part of the system, and the reserve has wolves, lynxes and snow leopards.

70 Wolong Nature Reserve, SOUTHERN
CHINA

The Wolong Nature Reserve is warm and wet, with misty mountains, ravinelike river valleys, and several forest zones climbing the slopes. Evergreen forests form the lowest zone, with a band of broadleaved trees above them, merging into a final swathe of conifers up to the tree line. The highest peaks are well over 19,700 feet (6,000 m), so there are considerable uplands of bare, rocky terrain. The torrential downpours of the summer monsoons turn earth to mud and loosen rocks already destabilized by frost and the region's frequent earthquakes. The resultant mud and rock avalanches leave livid scars in the mountain forests, which are soon recolonized by bamboo and shrubs.

Wolong's most famous denizen is the giant panda, and the reserve hosts China's largest population of this endangered species, with up to 70 individuals. The reserve also contains other rarities, such as the tree-climbing red panda, the clouded leopard and the golden monkey. Brilliant plumage is common among the reserve's birds, such as Temminck's tragopan and the Chinese monal.

71 Asir National Park SAUDI ARABIA

Saudi Arabia's southwestern Asir province flanks the Red Sea. Here, the Asir mountain range plunges to the ribbon of coastal plain known as the Tihamah. The 10,000-foot (3,000-m) scarp stands in the path of the northwest monsoons, and each year enough rain falls to convert the dry beds of the region's rivers into roaring torrents for a while. The Asir National Park contains mountains, juniper woodlands, humid scrub plains and bakingly hot beaches.

A land bridge once existed between Africa and Saudi Arabia, and many of the species in the park have African origins, including baboons, jackals, hyraxes and jerboas. Among the birds are African bearded vultures, Nile Valley sunbirds, gray hornbills and Bataleur eagles. The park is an important refuge in Saudi Arabia, where the popularity of hunting has pushed some species, such as the Arabian bustard, almost to extinction. The park is also on a major migratory route between Europe and eastern Africa used by over 200 species, including swallows, warblers and birds of prey.

72 Thar Desert NORTHWEST INDIA/PAKISTAN

Covering 270,270 sq. miles (700,000 km²) in the states of Rajasthan and Gujarat, the Thar Desert straddles the long northwestern border with Pakistan. The variety of desert habitats is large, with static and mobile dunes, rocky pavements and outcrops, and salt flats. The desert is not plantless, and there are large areas of seewan grass, as well as both wiry and succulent shrubs, and sparse trees.

Many of the desert creatures have extraordinary abilities to conserve water. The wild asses can hold reserves of water in their bodies for when it is needed and can survive extreme dehydration. The chinkara gazelle can go entirely without water, producing metabolic water from its food. The desert is not an attractive habitat for most humans, so the wildlife of the Thar is still diverse, with large numbers of birds and insects as well as mammals such as foxes and jackals. The Indian wolf has suffered at the hands of shepherds, having turned to sheep as prey when antelope numbers fell due to overhunting by humans.

73 Sagarmatha National Park NEPAL

Famous as the Everest park, Sagarmatha contains a stupendous collection of peaks, with seven summits over 23,000 feet (7,000 m), including the 29,078-foot (8,863-m) Everest. This is the park at the top of the world. It is situated in northeastern Nepal, and its northern boundary coincides with the Tibetan border. The Dudh Kosi River, and its tributaries, which eventually join up with the Ganges system, draws its headwaters from the park's four main glaciers via deeply carved valleys.

With its lower boundary at 9,334 feet (2,845 m), Sagarmatha contains relatively few mammals. Those that do inhabit the park include Himalayan musk deer, Asiatic black bears, snow leopards and red pandas, as well as wooly hares, Tibetan water shrews and short-tailed moles. An important high altitude breeding ground for birds, Sagarmatha contains 36 species which breed in the park, such as blood pheasants, Tibetan snowcocks and Himalayan monals. To offset the effects of tourism, special no-go areas are being established where all human activity is forbidden.

74 Kaziranga National Park ASSAM,
NORTHEASTERN INDIA

Kaziranga lies on the banks of the Brahmaputra River, and once a year, as the monsoons swell the waters to a raging flood, wildlife and people have to escape to high ground. The retreating river leaves extensive shallow swamps behind it, and these characterize the park, together with flat plains of elephant grass, and patches of evergreen forest. Kaziranga has been a reserve since 1926, when it was established to protect its rhinos from extinction.

The forest patches and tall elephant grass provide ideal stalking country for the good-sized tiger population in Kaziranga. Prey animals are plentiful, but tourist visitors wisely travel by elephant to view the wildlife rather than run the risk of encountering tigers. In addition to the tame riding elephants, the park contains large herds of migrating wild elephants. Other common inhabitants include buffaloes, swamp deer, gaurs, leopard cats and otters. The park is also rich in birds of prey, with several species of fishing eagles commonly seen.

75 Xishuangbanna Nature Reserve
SOUTHEASTERN CHINA

The 770 sq. miles (2,000 km²) of the Xishuangbanna Nature Reserve in Yunnan province lie within a tropical rainforest and house an exceptional variety of plants, some 50 percent of which are on China's protected list. The forest of the reserve is dense and luxuriant, with lianas, parasitic plants and other creepers trailing from the high canopy. Some trees, like the lofty fig, a type of banyan, also have aerial roots which grow down from the branches.

The reserve is not all solid forest, and rivers create open corridors. There are also outcrops of limestone, with unusual narrow pinnacles, and caves. Yet the vigor of the rainforest never ceases in the search for more space, and trees grow even on the exposed rocks, clinging to fissures and hanging onto boulders with nets of encircling roots. Asian elephants, gaurs and Hoolock gibbons all live within the reserve, and many species of animals and birds may have originally evolved in the region.

76 Khao Yai National Park THAILAND

One of the Asian mainland's last unexploited tropical forests, the 837 sq. mile (2,168 km²) Khao Yai National Park, once a refuge for bandits, remains one of the few reserves where the visitor can see tigers, leopard cats, Malaysian sun bears and gibbons in the wild, using trails that were made by wild elephants rather than park rangers. The terrain is mountainous, and in fact Khao Yai is part of the Phanom Dongrek range, with three peaks over 4,000 feet (1,200 m) in the park, and an upland rainfall that averages 120 inches (3,000 mm). Despite the torrential rains of July to October, there is a dry season that begins in November, during which all but the very largest water courses dry out completely.

The varied forest gleams with flowers, as well as such birds as scarlet minivets, green magpies, blue-winged leaf-birds and vernal hanging parrots. Hornbills and mynas feed on the fruit of strangling fig trees, which are also a vital food source for many other forest creatures, including gibbons, macaques, palm civets and certain bats.

77 Wilpattu National Park SRI LANKA

Largest of Sri Lanka's reserves, Wilpattu covers 508 sq. miles (1,316 km²) in the northwest of the island and is bounded by the sea on its western side. All across the park are the shallow basinlike pans known as *villus*, built many centuries ago by local princes as a form of irrigation, for Sri Lanka has no natural lakes, and the water table is deep beneath the surface. The *villus* have become the focus for wildlife within the park. Habitats vary, with sand dunes, dry thornbush terrain and deep forested regions.

Wilpattu is famous for its leopards and also has a bear population. Bears can sometimes be seen devouring ants from the park's many anthills, especially after rain. No elephants are resident within park boundaries, but migrating herds of 30 or more pass through, avoiding human contact as much as possible, and hiding out in the forest during the day. In this dry terrain, the *villus* are an irresistible magnet to wildlife such as barking deer and mouse deer, which in turn attract leopards.

SOUTHEAST ASIA/AUSTRALASIA

78 Iglit Baco National Park PHILIPPINES

Population density and the lumber industry have destroyed much of the Philippines' wild forest land, but difficulty of terrain and a flood-ridden rainy season for half of the year have helped the Iglit Baco National Park in the center of the island of Mindoro to avoid the worst depredations. Several tribal communities live within the park, hunting and practicing shifting rice cultivation. The western region of the park is mainly grassland, while in the eastern portion, evergreen forest and dry-season deciduous trees and shrubs predominate.

The mountainous center of the park, dominated by the 8,163-foot (2,488-m) Mount Baco, contains extensive grasslands riven by steep canyons and ravines. The endemic tamarau, a miniature water buffalo-like herbivore only 3 feet (1 m) high, lives in the mountain grasslands as well as in swampy lowland areas. Other mammals include prickly porcupines, wild pigs, deer and monkeys. Endemic Philippine birds, such as the tarictic hornbill and the yellow-breasted fruit dove, share the park with parrots, doves, swifts and kingfishers.

79 Kinabalu National Park EAST MALAYSIA

The multiple granite peaks of Mount Kinabalu tower above the Kinabalu National Park in the Malaysian state of Sabah in northern Borneo, a few degrees north of the equator. Visitors to this luxuriously forested park, intent on climbing Borneo's highest mountain, pass through several distinct zones of forest growth on the ascent.

First, there is a lowland region of lumber trees, where the rare rafflesia, the world's largest flower, is sometimes found. Above 4,000 feet (1,200 m) is a zone of mixed montane forest, rich in oaks and conifers. Brilliant rhododendrons, and some of Kinabalu's thousand species of orchids grow here, alongside carnivorous pitcher plants. Tree ferns and climbing bamboos flourish in the mossy environment of the next zone of cloud forest, up to 10,830 feet (3,300 m), after which comes the open summit

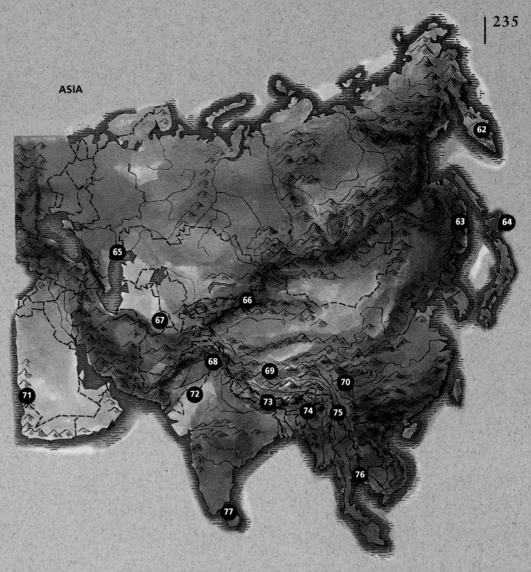

ASIA

region of granite slabs up to 13,455 feet (4,100 m). With a reputation as one of the best managed parks in the region, Kinabalu has over 300 bird species and 100 mammal species as well as its unique flora.

80 Mulu National Park SARAWAK, EAST MALAYSIA

Mulu is an extraordinary forest environment which can be entered only by means of long boat voyages up a series of rivers. The park has mountains, caves, alluvial forests, surreal landscapes of sandstone pinnacles, and a stupendous annual rainfall of between 25 and 200 inches (600 and 5,000 mm). But, there is a grave threat hanging over Mulu – that of the logging companies. They have already stripped out the lumber along the river banks. Penan tribespeople who live in the park barricaded logging roads and created international publicity in an attempt to halt wholesale logging in the late 1980s, but the loggers have strong political support, and the logging continues.

In the meantime, Mulu is still a wildlife enthusiast's dream, with almost every bird known to exist in Borneo, as well as several primates, such as macaques, leaf monkeys and Bornean gibbons. There is also a huge bat population centered in a number of caverns, and other mammals include sun bears, slow loris, western tarsiers and yellow-throated martens.

81 Mount Leuser National Park SUMATRA, INDONESIA

This great national park, covering 3,860 sq. miles (10,000 km²) of Indonesia's westernmost island, is of international importance as a major refuge of the

endangered Sumatran two-horned rhinoceros, and for its vital rehabilitation centers for orphaned orangutans. It has a long history of protection. The range of habitats includes coastal forest on Sumatra's west coast, mountain highlands which include the 11,093-foot (3,381-m) Mount Leuser, and extensive riverine terrains on both banks of the great River Alas which bisects the park.

The richness of Leuser's fauna is hard to exaggerate. In addition to the rare rhino, the forests conceal tigers, clouded leopards and barking deer, as well as primates including Thomas's leafmonkeys, long-tailed macaques, orangutans and siamang gibbons. The rare Sumatra elephant prefers the river valleys, where there are salt licks, and the valleys are also home to otters, hog badgers and leopard cats. In all, there are 176 mammal species, some 320 bird species, 194 reptile species and 52 amphibian species.

82 Komodo National Park LESSER SUNDAS, INDONESIA

The climate of the Lesser Sundas is relatively arid, with sparse tree cover on the uplands. The islands of the group lie in Australia's rain shadow and receive rain only between November and March. Komodo National Park consists of three islands, Komodo, Rinca and Padar, and is the home of the planet's heaviest lizard, the Komodo dragon.

Virtually unknown outside its home islands until the early 20th century, the Komodo dragon is a giant monitor lizard which can reach 10 feet (3 m) in length and weigh 330 pounds (150 kg). It preys on a variety of creatures, including deer, feral water buffaloes, pigs and horses, all of which thrive in the park's savanna woodland and open grassland. The park lies in an overlap area between Asia and Australia, with species typical of both continents, notably among the bird population. Australian species such as sulfur-crested cockatoos and friar birds share terrain with Asian jungle fowls and monarch flycatchers.

83 Central and Southern Highlands PAPUA NEW GUINEA
Despite the encroachment of mining and logging, Papua New Guinea still retains most of its rugged wild landscape, especially in the mountains. Below 10,000 feet (3,000 m) are mid-montane forests, with animals and plants that have adapted to survive only here and can live neither higher up nor lower down. In the misty, overcast environment live several of the bird-of-paradise species in which the island is so rich. The plant life includes great beech trees closely related to beeches in New Zealand and South America, survivors from the breakup of ancient Gondwana. Brilliant color among the trees comes from mosses, fungi and many of New Guinea's 2,500 species of orchids.

Above 10,000 feet (3,000 m), the upper montane begins, with thick banks of wet moss everywhere. Bowerbird species build their bowers up here, decorated with shells and feathers, and canopies of orchid stems. Above the moss forest is a zone of subalpine grassland and stunted woodland, where wild dogs and spotted marsh harriers hunt small rodents and mountain wallabies graze.

84 Kakadu National Park AUSTRALIA
Aboriginal peoples have inhabited Arnhem Land in Australia's Northern Territory for over 30,000 years, and Aboriginal organizations own half of the area of the Kakadu National Park on the Arnhem coast, taking part in management decisions for Australia's largest and most diverse preserve. The park incorporates freshwater marshes, tidal estuaries, salt flats and mangrove swamps on offshore islands.

Kakadu teems with birdlife, dominated in peak periods by 8-foot (2.5-m) magpie geese. Other wetland inhabitants include huge saltwater crocodiles. In the dry season, saltwater tides penetrate the flood plain of the rivers that cross and define the park, carrying saltwater fish, including sharks, up to 20 miles (30 km) inland. Freshwater crocodiles and long-necked turtles inhabit creeks and ponds.

85 Great Barrier Reef AUSTRALIA
The largest system of coral reefs in the world, the Great Barrier Reef stretches for more than 1,250 miles (2,000 km) down the Queensland coast, a vast complex of 2,900 separate reefs and over 500 islands and islets. Growing threats from tourism, the oil industry and agricultural effluvia led in 1975 to the creation of the Great Barrier Reef Marine National Park, covering 98.5 percent of the reef's area.

The reef is a huge and complex ecosystem, containing 400 species of corals, which have suffered serious inroads from population explosions of the crown-of-thorns starfish in recent years. Seagrass beds shelter large populations of dugong, which, along with all sea turtles, are fully protected within the park. Over 240 species of birds nest on the coral cays, including terns, white-bellied sea eagles, reef herons and ospreys. At least 1,500 species of fish and 4,000 species of mollusks help make the reef one of the world's most important marine biospheres.

86 The Olgas CENTRAL AUSTRALIA
Formed from a conglomerate of granite, gneiss and volcanic rocks, and resisting the erosion that over millions of years pared back the plain around them,

the Olgas are a group of some 30 massive rocky domes covering an area of 13½ sq. miles (35 km²), several of which are over 1,650 feet (500 m) in height. In the steep clefts and chasms dividing the red domes are shaded rock pools, oases of life festooned with vines and sheltering lizards, snakes, birds and small mammals. Gum trees grow in the creek beds in some of the ravines, alongside mosses, ferns, yellow-flowered cassia and acacia bushes.

The sheer, curving sides of the individual domes are bare, though some are crowned with spinifex. The Olgas, like nearby Uluru (Ayer's Rock), are a sacred site of the Aborigines, who believe a sacred serpent sleeps in a cavern inside Mount Olga, the largest of the domes. The serpent's breath is the wind that sometimes howls through the gorges.

87 The Simpson Desert CENTRAL AUSTRALIA
The great red dunes of the Simpson Desert run in parallel lines, some as much as 75 miles (120 km) long. More than 1,000 of these dune ridges range across the desert, up to 130 feet (40 m) high, and with corridors 330 feet to ½ mile (100 m to 1 km) wide between them. Covering over 56,000 sq. miles (145,000 km²), it is one of the driest regions of Australia.

With savage daytime temperatures and little shade, the Simpson hosts specially adapted populations which leave their tracks in the sand overnight. The sand goanna preys on small marsupial mammals, insects, scorpions, birds and other reptiles. The most common vegetation is spinifex, growing in sparse and spiny clumps. After the rare rain showers, desert flowers bloom profusely, then die quickly, leaving their seeds lying dormant until the next rain. Some birds, like the black-faced wood swallow, begin their breeding cycles the moment it rains.

88 Lake Eyre CENTRAL AUSTRALIA
South of the Simpson Desert, Lake Eyre is situated on top of the planet's largest artesian system, the Lake Eyre Basin, draining an area of 425,000 sq. miles (1.1 million km²). Four major rivers drain into the lake, yet it is arid and apparently waterless most of

the time, forming a huge salt flat of about 2,320 sq. miles (6,000 km²), most of which lies 50 feet (15 m) below sea level. Creatures living in this glaring blast furnace have extraordinary survival techniques.

The eggs of the brine shrimp can remain for years in suspended animation until rain starts their brief life cycle. The lizards known as Lake Eyre dragons hibernate in burrows in the salt crust and use their eyelids as sun visors. A few times a century, an errant tropical cyclone dumps quantities of water into the lake's huge drainage basin, flooding vast areas of salt flat. In 1974, the lake experienced its biggest floods in perhaps 500 years, covering an area of 350 sq. miles (900 km²) to a depth of 13–16 feet (4–5m). During the rare floods, many birds, including cockatoos, ducks, waders and pelicans, gravitate to the lake from all over the continent.

89 The Snares Islands NEW ZEALAND
Some 125 miles (200 km) southwest of New Zealand's South Island lie the Snares, a group of small islands. The two largest in the group are North East Island, of 692 acres (280 ha), and Broughton Island, of 119 acres (48 ha). Protected by wild and rocky coasts and formidable seas, the Snares are an important bird sanctuary. A forest of *Olearia*, or giant tree daisy, and tussock meadows cover much of the main islands, which have floors of black peat riddled with the burrows of some six million sooty shearwaters.

The islands have three unique species of landbird: the Snares black tit, the Snares fernbird and the Snares snipe. Other landbirds include silver-eyes, redpolls, blackbirds, thrushes and gray warblers. The seabirds dominate the islands' fauna, with a large population of 80,000 Snares crested penguins. New Zealand fur seals live on the main islands and the islets of the Western Chain. A small bachelor herd of Hooker's sea lions visits annually, as do a few elephant seals. The Snares have never suffered the introduction of species associated with humans, such as rats and cats, and have thus avoided the destruction of native species that such introductions have brought to many islands.

SOUTHEAST ASIA AND AUSTRALASIA

Photographic credits

l = left, *r* = right, *t* = top,
c = center, *b* = bottom

1 Eric Lawrie/Royal Geographical Society; 2/3 Thomas Kitchin/Tom Stack & Associates; 4 Paul McCormick/The Image Bank; 5 Anthony Bannister/NHPA; 6 Peter Davey/Bruce Coleman; 6/7 Hans Christian Heap/Planet Earth Pictures; 7 Ben Osborne/Oxford Scientific Films; 10/12 Nikita Ovsyanikov/Planet Earth Pictures; 13*t* Daniel Cox/Oxford Scientific Films; 13*b* Francisco Erize/Bruce Coleman; 14/15 Bryan & Cherry Alexander; 16 Jim Brandenburg/Zefa Picture Library; 17 Jim Brandenburg/Planet Earth Pictures; 18/19 Jim Brandenburg/Zefa Picture Library; 19*t* Jim Brandenburg/NHPA; 19*b* Jim Brandenburg/Zefa Picture Library; 20/21 Jill Ranford/Ffotograff; 22 Martin W. Grosnick/Ardea; 23 Richard & Julia Kemp/Survival Anglia; 24/25 D. Parer & E. Parer-Cook/Ardea; 26/27 Ben Osborne/Oxford Scientific Films; 28/29 Colin Monteath/Mountain Camera; 30/31 J.E. Pasquier/Rapho; 32 Paul McCormick/The Image Bank; 33*t* John Shaw/Tom Stack & Associates; 33*b* John Eastcott/Planet Earth Pictures; 34 Steve Kaufman/Peter Arnold; 34/35 Douglas T. Cheeseman Jr./Peter Arnold; 36/37 Stephen Krasemann/NHPA; 38/39 Julian Pottage/Robert Harding Picture Library; 41*t* Patti Murray/Oxford Scientific Films; 41*b* Survival Anglia; 42*l* Erwin & Peggy Bauer/Bruce Coleman; 42/43 Daniel J. Cox/Oxford Scientific Films; 44/45 Tom Mangelsen; 46 Tom Ulrich/Oxford Scientific Films; 47*t* Tom Mangelsen; 47*b* Frank Schneidermeyer/Oxford Scientific Films; 48/49 David Muench; 50 Jen & Des Bartlett/Bruce Coleman; 51*t* M.P. Kahl/Bruce Coleman; 51*b* Charlie Ott/Bruce Coleman; 52/53 David Muench; 54/55 David Muench; 55/56 John Shaw/NHPA; 57 David A. Ponton/Planet Earth Pictures; 58/59 Dieter & Mary Plage/Bruce Coleman; 60*t* Kevin Schafer/NHPA; 60*b*/61 Michael Fogden/Oxford Scientific Films; 62/63 Ivor Edmonds/Planet Earth Pictures; 64 Chris Prior/Planet Earth Pictures; 65 James H. Carmichael/The Image Bank; 66/67 Tony Morrison/South American Pictures; 69 Kimball Morrison/South American Pictures; 71 Adrian Warren/Ardea; 72/73 André Bärtschi/Planet Earth Pictures; 74 Stephen Dalton/NHPA; 75*t* K.W. Fink/Ardea; 75*b* André Bärtschi/Planet Earth Pictures; 76 Partridge Productions/Oxford

Scientific Films; 77 Kevin Schafer/Tom Stack & Associates; 78 Robin Hanbury-Tenison/Robert Harding Picture Library; 78/79 Thomas Kelly/Impact; 79 Victor Englebert/Select Photo Agency; 80/81 Carlos G.E. Velha/The Image Bank; 82*t* Luiz Claudio Marigo/Bruce Coleman; 82*b* George Gainsburgh/NHPA; 83 Richard Matthews/Planet Earth Pictures; 84/85 Eric Lawrie/Royal Geographical Society; 86*t* Gunter Ziesler/Bruce Coleman; 86*b* François Gohier/Ardea; 87 Roger Few; 88/89 Tony Morrison/South American Pictures; 89*t* Tony Morrison/South American Pictures; 89*b* John Bulmer/Comstock; 90/92 Richard Matthews/Planet Earth Pictures; 93*t* Richard Coomber/Planet Earth Pictures; 93*b* Richard Matthews/Planet Earth Pictures; 94/95 Bernadette Waters/Planet Earth Pictures; 96 Julie Bergada/Aspect Picture Library; 97 Konrad Wothe/Oxford Scientific Films; 98/99 Boireau/Rapho; 100 Mike Brown/Oxford Scientific Films; 101*l* Boireau/Rapho; 101*r* G.K. Brown/Ardea; 102/104*l* Keith Scholey/Planet Earth Pictures; 104/105 Owen Newman/Oxford Scientific Films; 106/107 Jonathan Scott/Planet Earth Pictures; 108 Stephen Krasemann/Peter Arnold; 108/109 Rafi Ben-Shahar/Oxford Scientific Films; 110/111 Peter Davey/Bruce Coleman; 112 Mirella Ricciardi/Colorific!; 112/113*t* Brian Boyd/Colorific!; 112/113*b* Jonathan Scott/Planet Earth Pictures; 114/115 John Newby/WWF International; 115 Anthony Bannister/Oxford Scientific Films; 116 Nick Gordon/Ardea; 117 Jacques Jangoux; 118/119 Anthony Bannister/NHPA; 120 Carol Farneti/Planet Earth Pictures; 121 Mike Rosenberg/Oxford Scientific Films; 122 Richard Packwood/Oxford Scientific Films; 122/123 E.A. Janes/NHPA; 123 Carol Farneti/Planet Earth Pictures; 124 Anthony Bannister/NHPA; 124/125 Tom Nebbia/Aspect Picture Library; 125 Nigel Dennis/NHPA; 126/127 Peter Johnson/NHPA; 128 Anthony Bannister/Oxford Scientific Films; 129 Antoinette Jaunet/Aspect Picture Library; 130 Anthony Bannister/NHPA; 131*t* David Hughes/Bruce Coleman; 131*b* Michael Fogden/Oxford Scientific Films; 132/133 Bengt Olof Olsson/Bildhuset; 134 Ferrero/Labat/Ardea; 135*t* Johnny Johnson/Bruce Coleman; 135*b* John Noble/Wilderness Photographic Library; 136 Pelle Stackman/Tiofoto AB; 137 Robert Harding Picture Library; 138/141 David Paterson; 142/143 L. Campbell/Scotland in

Focus; 143*t* Alan & Sandy Carey/Oxford Scientific Films; 143*b* Gordon Langsbury/Bruce Coleman; 144/145 Bomford & Borkowski/Survival Anglia; 146/147*l* David Woodfall/NHPA; 147*r* Hans Reinhard/Zefa Picture Library; 148/149 Ivor Edmonds/Planet Earth Pictures; 150 David Hughes/Robert Harding Picture Library; 151*t* Nigel Dennis/NHPA; 151*b* Hervé Berthoule/Explorer; 152/153 Francisco J. Erize/Bruce Coleman; 154 Jose Luis Gonzalez Grande/Bruce Coleman; 155*t* Ian Beames/Ardea; 155*b* Robert Frerck/Robert Harding Picture Library; 156/157 Comstock; 158/159 Hans Christian Heap/Planet Earth Pictures; 160/161*t* Comstock; 161*c* Robin Constable/Hutchison Library; 161*b* Alan Keohane/Impact; 162/164*t* Joel Bennett/Survival Anglia; 164*b* Michael Leach/NHPA; 165 Joel Bennett/Survival Anglia; 166/167 Tony Allen/Oxford Scientific Films; 168/169 Richard Kirby; 170 Doug Allan/Oxford Scientific Films; 170/171 John Hartley/NHPA; 172/173 Sarah Leen/Matrix/Colorific!; 174/175 Colin Monteath/Mountain Camera; 176*t* Tim Davis/Oxford Scientific Films; 176*b*/177 Martyn Colbeck/Oxford Scientific Films; 178/179 Mike Price/Bruce Coleman; 180 Paolo Koch/Okapia; 181 Rod Williams/Bruce Coleman; 182/183 Ann & Bury Peerless; 183 Duncan Maxwell/Robert Harding Picture Library; 184 Rod Williams/Bruce Coleman; 185*t* Dieter & Mary Plage/Survival Anglia; 185*b*/186*l* Carlo Dani & Ingrid Jeske/Natural Science Photos; 186/187 Anup Shah/Planet Earth Pictures; 187 Gunter Ziesler/Bruce Coleman; 188/189 Gerald Cubitt; 190 D. Parer & E. Parer-Cook/Auscape; 191*l* Alain Compost/Bruce Coleman; 191*r* Gerald Cubitt; 192 Bryan & Cherry Alexander; 192/193 Adrian Arbib/Royal Geographical Society; 193*l* J. Riley/Hedgehog House; 193*r* Robert Harding Picture Library; 194/195 Comstock; 196/199 Dieter & Mary Plage/Survival Anglia; 200/201 Sam Abell/National Geographic Society; 202 Robert Harding Picture Library; 202 John Lythgoe/Planet Earth Pictures; 203 Jean-Paul Ferrero/Auscape; 205*t* Reg Morrison/Auscape; 205*b* Jan Taylor/NHPA; 206 Purdy & Matthews/Planet Earth Pictures; 206/207 D. Parer & E. Parer-Cook/Ardea; 207 Purdy & Matthews/Planet Earth Pictures; 208/209 Natural Images/NHPA; 210 Belinda Wright/Oxford Scientific Films; 211 Jean-Paul Ferrero/Auscape; 212/213 Richard Packwood/

Oxford Scientific Films; 214*t* John Cancalosi/Bruce Coleman; 214*b*/215 Jean-Paul Ferrero/Ardea; 216/217 Darryl Torckler/Hedgehog House; 218 G.R. Roberts; 219*t* Darryl Torckler/Hedgehog House; 219*b* Kim Westerskou/Hedgehog House; 220/221 Superstock; 222*t* Frances Furlong/Survival Anglia; 222*b* Norman Tomalin/Bruce Coleman; 224 Tom Till/Auscape; 225*l* Tom Ulrich/Oxford Scientific Films; 225*r* John Cancalosi/Bruce Coleman

Illustration credits

Maps throughout: **Andrew Farmer**

Diagrams: **Gary Hincks, Janos Marffy, Richard Bonson and Andrew Farmer**

Editorial director **Ruth Binney**
Managing editor **Lindsay McTeague**
Text editor **Isabella Raeburn**
DTP **Pennie Jelliff**
Production **Sarah Hinks**
Kate Waghorn
Coordinator **Tim Probart**

Typeset by **Millions Design**
Origination by **CLG**, Verona, Italy

If the publishers have unwittingly infringed copyright in any illustration reproduced, they would pay an appropriate fee on being satisfied to the owner's title.